URBAN RENEWAL, ETHNICITY AND SOCIAL EXCLUSION IN EUROPE

Urban Renewal, Ethnicity and Social Exclusion in Europe

Edited by

ABDUL KHAKEE
Centre for Housing and Urban Research
University of Örebro

PAOLA SOMMA
DAEST
University of Venice

HUW THOMAS
Department of City and Regional Planning
Cardiff University

Ashgate

Aldershot • Brookfield USA • Singapore • Sydney

Published by
Ashgate Publishing Ltd
Gower House
Croft Road
Aldershot
Hants GU11 3HR
England

Ashgate Publishing Company
Old Post Road
Brookfield
Vermont 05036
USA

British Library Cataloguing in Publication Data
Urban renewal, ethnicity and social exclusion in Europe
1. Marginality, Social - Europe, Western 2. Minorities - Europe, Western - Social conditions 3. Racism - Europe, Western 4. Urban renewal - Social aspects - Europe, Western
I. Khakee, A. (Abdul) II. Somma, Paola III. Thomas, Huw, 1954-
305.5'6'094

Library of Congress Catalog Card Number: 98-74197

ISBN 1 85972 664 X

Printed and bound by Athenaeum Press, Ltd.,
Gateshead, Tyne & Wear.

Contents

Tables and figures

Portugal

Britain

Contributors

Rabia Bekkar
IPRAUS (CNRS), and
University of Paris X Nanterre,
France

Marcus Johansson
Centre for Housing and Urban Research
University of Örebro
Sweden

Abdul Khakee
Centre for Housing and Urban Research
University of Örebro
Sweden

Ton van der Pennen
Sociaal en Cultureel Planbureau,
Rijkswijk ZH,
The Netherlands

Daniel Pinson
Institut d'Aménagement Régional d'Aix en Provence,
University Aix-Marseilles III,
France

Carlos Nunes Silva
Centro de Estudos Geográficos,
University of Lisbon,
Portugal

Vincent Smit
VROM-advisory Board,
The Hague,
The Netherlands

Paola Somma
DAEST
University of Venice
Italy

Huw Thomas
Department of City and Regional Planning,
Cardiff University,
Wales, UK

Jan Willem van de Wardt
BRON UvA B.V.,
University of Amsterdam,
The Netherlands

Acknowledgements

This book took far longer to produce than was first envisaged, as contributors juggled various commitments. Their patience and hard work helped ensure a successful outcome. The efforts of Anne Scales were vital in securing the book's production. It was she who rigorously copy-edited the manuscript and prepared it as a camera-ready document. Without her energy and attention to detail the book would not have been published.

1 Introduction

ABDUL KHAKEE, PAOLA SOMMA AND HUW THOMAS

Background

The background to the concerns explored in this book has two main elements. Firstly, that racism is widespread, and deeply rooted, in Western Europe, and that every country provides evidence of minorities whose economic and social lives are seriously affected because of discriminatory treatment (Solomos and Wrench, 1993). Secondly, that the way in which immigration has been politicised means that the continuing pressure to accept migrants to Western European countries (Salt, 1996) will fuel racism and will very likely shape the terms in which racial discrimination is debated (see, e.g., Solomos, 1993, on the UK experience, and Demker, 1993, on the Swedish experience). This has implications for all racialised minorities, not simply immigrants. Adopting the terminology which has become fashionable in policy circles in Europe, the book talks of the 'exclusion' of minorities. We have reservations about some uses of this terminology, but hope its use will neither confuse readers nor detract from the merits of contributions to this volume. The book is a stocktaking of our current knowledge of the impact of urban renewal policies on the social exclusion of ethnic minorities in six European countries. Social exclusion has received considerable attention from policy-makers and researchers in recent years (see, e.g., Brown and Crompton, 1994; Roche and van Berkel, 1997; Room, 1995a; International Planning Studies, 1998). Yet *spatial* aspects of exclusion have received comparatively little attention. This book cannot hope to redress this imbalance completely, but it marks a starting point.

There are two main reasons why the relationship between urban renewal and social exclusion merits closer and more systematic study. First, the study of urban renewal episodes can provide a useful insight into the dynamics of exclusion. In particular, if care is taken in the definition of what constitutes 'urban renewal', then its study might prove a useful focus for comparative research, given the ubiquity of spatial restructuring. This ubiquity brings us to the second point, namely that the spatial restructuring of towns and cities is bound up with contemporary processes of global economic change, and the implications of the latter for social exclusion cannot be fully understood without spatial analysis.

The two concepts underlying this study - urban renewal and social exclusion - are intricate and complex. As the case studies in this volume show, *context* plays a substantial role in determining the policy contents and implications of these two concepts. The term 'urban renewal' is a difficult one, partly because, in a generic sense, it encompasses various types of intervention to tackle the problems of older parts of towns and cities, including their central business districts (Gibson and Langstaff, 1982). What constitutes 'older parts' varies from country to country depending on each country's specific history and pattern of urbanisation. The nature of intervention has also changed owing to the changing perception of the problems of inner cities and alleged 'urban ghettoes'. This change has been explained in different ways. For example, according to Atkinson and Moon (1994), inner cities' problems were until the mid 1970s explained in terms of 'the pathological behaviour of their residents', but by the late 1970s there was an increasing awareness of the 'wider structural nature of the problem'. The latter also explains why the term 'urban renewal' has been replaced by 'urban regeneration' in urban policy literature. It has meant a shift, from a concern with property and land, to social and economic concerns. Gibson and Langstaff (1982) distinguish five phases in the development of urban renewal policies in the UK: slum clearance and redevelopment, housing and environmental improvement, gradual renewal combining selective clearance with improvement, experiment-based eradication of urban deprivation, and finally, a comprehensive approach incorporating economic and social improvement. Similar shifts in policy emphasis can be discerned in other Western European countries.

The term 'social exclusion' is - despite (or perhaps because of) its ubiquity - a term without a clear and unambiguous meaning. Room (1995b) has argued that a concern with exclusion fits most comfortably into a view of society as normally (or ideally) exhibiting a high degree of integration, or cohesion, exclusion from which may be deemed a social ill.

Social exclusion is defined by Berghman (1995, p. 19) as:

> ... failure of one or more of the following systems:
>
> - the democratic and legal system, which promotes civic integration
> - the labour market, which promotes economic integration
> - the welfare state system, promoting what might be called social integration
> - the family and community system, which promotes interpersonal integration.

Berghman's definition has the merit of identifying a number of dimensions of social exclusion, and it is this multi-dimensionality that helps to distinguish even a careful and modest interpretation of the concept from the notion of poverty. Yet the dimensions must not be conceived as sealed spheres of social life, nor as generating independent criteria, to be checked off in some social exclusion audit. They are interdependent and interpenetrating. A second observation is that most discussion of social exclusion assumes that overcoming it is a 'good thing'. This is most clear in relation to definitions such as that of Berghman. As Room (1995b) has pointed out, one of the most prominent intellectual currents within which a concern for social exclusion makes sense is a conservative continental European one, which identifies as a serious social ill any detachment from the moral order which binds, as a single society, a group of collectivities, with defined interactive mutual obligations and claims. We may question whether this view of society is either plausible or coherent, but if (for the sake of argument) it is granted some credence, it is clear that too focused a concern for minimising detachment from the moral order can leave unquestioned the possibility of the order's being oppressive for certain groups (Young, 1990). Alden and Thomas (1998) suggest a more modest understanding of social exclusion, which contrasts it with *participation* rather than *integration*. Exclusion, on this account, is the result of the existence of barriers to identifying, or taking up, *opportunities* to participate in economic, political and social life. While contributors to this book were not asked to explicitly discuss conceptions of social exclusion, it is the latter conception which is implicit in their analyses.

Relationship between urban renewal and social exclusion

Contributors to the volume were asked to review current understanding of the links between urban renewal and the social exclusion of ethnic minorities, focusing on three key processes:

- the use of urban policy frames;
- power relations within policy making processes;
- racialisation.

What constitutes an 'urban policy frame' is by no means straightforward. It is a complex and ambiguous phenomenon. There are a large number of theoretical approaches to the understanding of the evolution of urban policy frames in different countries. Atkinson and Moon (1994) outline three analytical traditions - pluralist, Marxist and neo-liberal - in order to analyse and evaluate urban policy in the UK. Underlying these three approaches are differences in the nature of interaction between developers, property speculators, other businesses, central and local government, city residents, and other organisations. In response to, and as a critique of, pluralist and neoliberal theories, regime theory has been increasingly used. This underscores the reliance of government on business in capitalist democracies (Lauria and Whelan, 1995). The Marxist tradition has been further developed, and claims to understand the interplay between market mechanisms and regulatory authorities which is required to secure stable capitalist accumulation (Feldman, 1995).

Given these approaches, we have defined 'policy frames' as ways of integrating 'facts, values, theories and interests' (Solesbury, 1993) and have suggested certain key dimensions of such policy frames. They prescribe:

- the relation of the state to the market;

- the relations between different tiers of the state;
- the role of urban governance.

The second key idea is that of power relations within policy making processes - at both the national level (where they influence the content of national urban policy frames) and the local level (where decisions on the detailed content and implementation of urban policy are made). An emphasis on the policy process raises a large number of interlocking issues with regard to agenda setting (how problems are defined and their causes recognised), development of policy measures and strategies (how goal-focused sets of measures are coherently presented), policy management (how organisational structures are co-ordinated to administer policy implementation), provision of political support and legislative mechanisms (how political powers and legislation facilitate efficient delivery of policy outputs), and evaluation and monitoring (in order to ensure that policy has achieved its aims, and that the views of those affected by a policy are fed back into the policy process) (see, e.g., Kingdon, 1984; Ham and Hill, 1985).

The objective is not to cover the entire policy process for urban renewal in order to demonstrate consequences in the form of social exclusion. Key questions to be addressed in discussing power relations include who has access to (or a voice in) policy making (technical personnel, sectors of the 'public', politicians, etc.); and how are power and influence mobilised. Answering the latter question involves considering the resources which are drawn upon, distinctive modes of discourse which are used, and the evaluative criteria deployed, to create a privileged position for certain groups within decision-making processes.

The final concept, racialisation, refers to the socially constructed nature of racial categories, and the contingent (and spatially variable) way in which these categories can penetrate and structure social relations, including, of course, relations of power. Racialisation consists of two measures. First, certain groups are ascribed certain phenotypical hereditary human characteristics. In the next phase, further negative attributes are ascribed to these groups (see, e.g., Miles, 1989; Hall and Du Gay, 1996). The object of racialisation is not the individual but specific groups' ways of living, often referred to using the notion of 'culture'. Amin (1989) states that the racist category 'white' has been replaced by the cultural category 'European'. It is no longer a question of superior race but a superior culture, 'the western culture' (see, e.g., Löwander, 1997).

The intensification of national and cultural forms of racism in Europe has been variously described as the 'crisis of the national state' (Miles, 1994), 'The modernity crisis' (Wieviorka, 1995), or 'xenophobia' (Hobsbawn, 1994). It is a result of increasing unemployment and insecurity in the labour market, and of the wearing away of the institutional and welfare networks which previously held people together. The kinds of racial distinction which are drawn, and their socio-political significance, varies from country to country, reflecting their histories (recent and more distant), and contemporary social and economic pressures. The nature of racialisation will influence the way in which urban policy is both conceived and implemented; i.e. it will influence policy frames and will structure the policy-making process.

In addressing the historically specific interrelationships of urban renewal, racialisation, and aspects of social exclusion, in various countries, each chapter follows a standard format, dealing with:

(i) institutions of local (municipal and regional) politics and government, and inclusion/exclusion of black and ethnic minorities;

(ii) the history of the black/ethnic minority population and the racialisation of politics - providing a national picture, but one sensitive to spatial variations;

(iii) urban policy frames, and their evolution in recent decades; i.e. normative perspectives within which urban policies (housing, urban renewal, etc.) are defined and evaluated;

(iv) the history of urban renewal (during the post 1945 period); and present trends;

(v) a review of evidence about the implications of urban renewal activity on the social exclusion of black and ethnic minorities;

(vi) selected case studies as appropriate;

(vii) an agenda for future research in the field of urban renewal and social exclusion.

The result, we believe, is a book which demonstrates the potential for comparative study in this area. We have produced a volume which can be categorised as a systematic compendium; a collection of some national case studies, which invites comparative research and analysis, but does not, in itself, guide it.

The value of comparative study

It has been persuasively argued that rigorous and fruitful cross-national comparative analysis of socio-economic processes, and of policy responses to them, is only possible within a theoretical framework which:

(i) identifies the specific ways in which national differences may influence the phenomena being studied (and thereby make a cross-national comparison worthwhile);

(ii) allows analysis to get beyond simply noting the inevitable institutional and administrative variety across states, by characterising significant socio-economic and political processes in general terms, and relating them (as necessary) to specific national institutional expression.

Thus Stoker and Mossberger (1994, p. 196) comment that 'concepts are the very foundation of comparison', and go on to quote Rose (1991, p. 447) to the effect that:

> Concepts are necessary as common points of reference for grouping phenomena that are differentiated geographically and often linguistically .... Methodologically, comparison is distinguished by its use of concepts that are applicable in more than one country.

Similarly, Allum (1995), in comparing politics in four western European states, does not organise his account around institutions but around the three key *concepts* of state, economy, and civil society. The suggestion arising from this book is that the three concepts identified earlier provide a useful framework for comparative study. Each concept identifies a social process which is mediated by 'national filters', a phrase which signifies the influence of distinctive national histories and distinctive national policy responses to the processes concerned. Collectively, the dynamics of the process delineated can create and reinforce social exclusion, but the precise manner in which this occurs is likely to vary across national boundaries. Part of the book's purpose is to argue that cross-national research is necessary to understand:

(i) the complexity of the pictures across at least a part of the EU, and, in particular, to understand the distinction between those 'national filters' which *are* significant in creating distinctive mechanisms for social exclusion, and those which are not;

(ii) the most fruitful and cost-effective way for policy intervention at EU level to operate, given that its influence, too, will be mediated by 'national filters'.

It represents no more than a contribution to a continuing policy debate, but, we hope, a useful contribution.

References

Alden, J. and Thomas, H. (1998), 'Social exclusion, space and labour markets', *International Planning Studies,* vol. 3, no. 1, pp. 7-13.

Allum, P. (1995), *State and Society in Western Europe,* Polity Press, Cambridge.

Amin, S. (1989), *Eurocentrism*, Zed, London.

Atkinson, R. and Moon, G. (1994), *Urban Policy in Britain: The City, the State and the Market*, Macmillan, London.

Berghman, J. (1995), 'Social Exclusion in Europe: Policy context and analytical framework', in Room, G. (ed.), *Beyond the Threshold: the measurement and analysis of social exclusion,* Policy Press, Bristol.

Brown, P. and Crompton, R. (eds) (1994), *A New Europe? Economic Restructuring and Social Exclusion,* UCL Press, London.

Feldman, M. (1995), 'Regime and Regulation in Substantive Planning Theory', *Planning Theory*, no. 14, pp. 65-94.

Gibson, M.S. and Langstaff, M.J. (1982), *An Introduction to Urban Renewal*, Hutchinson, London.

Hall, S. and Du Gay, P. (1996), *Questions of Cultural Identity*, Sage, London.

Ham, C. and Hill, M. (1985), *The Policy Process in the Modern Capitalist State*, Harvester Wheatsheaf, Brighton.

Hobsbawn, E. (1994), *Age of Extremes: The Short Twentieth Century 1914-1991*, Michael Joseph, London.

International Planning Studies (1998), 'Social exclusion in Europe', Special Issue vol. 3, no. 1.
Kingdon, J.W. (1984), *Agendas, Alternatives and Public Policies*, Little Brown, Boston.
Lauria, M. and Whelan, R.K. (1995), 'Planning Theory and Political Economy: the need for reintegration', *Planning Theory*, no. 14, pp. 8-33.
Löwander, B. (1997), *Racism and Anti-racism on the Agenda. Studies of Swedish Television News in the Beginning of the 90s*, Faculty of Social Science, Umeå University, Umeå (Ph.D. dissertation in Swedish with a summary in English).
Miles, R. (1989), *Racism*, Routledge, London.
Miles, R. (1994), 'Explaining racism in contemporary Europe' in Rattansi A. and Westwood, S. (eds), *Racism, Modernity, Identity on the Western Front*, Polity Press, Cambridge.
Roche, M. and van Berkel, R. (eds) (1997), *European Citizenship and Social Exclusion*, Ashgate, Aldershot.
Room, G. (ed.) (1995a), *Beyond the Threshold: the measurement and analysis of social exclusion*, Policy Press, Bristol.
Room, G. (1995b), 'Poverty and Social Exclusion: Two New European Agendas for Policy and Research' in Room, G. (ed.), *Beyond the Threshold: the measurement and analysis of social exclusion,* Policy Press, Bristol.
Rose, R. (1991), 'Competing forms of comparative analysis', *Political Studies*, vol. 39, pp. 446-462.
Salt, J. (1996), 'Migration Pressures on Western Europe' in Coleman, D. (ed.), *Europe's Population in the 1990s*, OUP, Oxford.
Solesbury, W. (1993), 'Reframing Urban Policy', *Policy and Politics*, vol. 21, no. 1, pp. 31-38.
Solomos, J. (1993), *Race and Racism in Britain*, 2nd ed., Macmillan, London.
Solomos, J. and Wrench, J. (eds) (1993), *Racism and Migration in Western Europe*, Berg, Oxford.
Stoker, G. and Mossberger, K. (1994), 'Urban regime theory in comparative perspective', *Environment and Planning C. Government and Policy*, vol. 12, no. 2, pp. 195-212.
Wieviorka, M. (1995), *The Arena of Racism*, Sage, London.
Young, I. M. (1990), *Justice and The Politics of Difference*, Princeton University Press, Princeton, NJ.

2 Not on our doorstep: Immigrants and 'blackheads' in Sweden's urban development

ABDUL KHAKEE AND MARCUS JOHANSSON

Introduction

The title of this paper needs explaining. As a result of public pressure, the Social Democrat Government of Sweden in 1996 declared that in future the country would be willing to pay remuneration to a 'third country' which was prepared to take care of political refugees that Sweden is obliged to accept according to the quotas prescribed by UNHCR. This measure is part of the policy which the Swedish Government has adopted in order to stifle immigration of refugees to the country, and to expel refugees who gave false information about themselves when seeking political asylum, or who committed petty crimes such as stealing the equivalent of £10 (Guillou, 1996). 'Not on our doorstep' is a telling description of the current immigration policy.

At the end of 1992, about 1.6 million persons in Sweden were officially designated as immigrants ('invandrare'). Of these, nearly 700,000 persons were born in Sweden but have at least one immigrant parent. The remaining 900,000 were almost equally divided into three categories: immigrants from other Nordic countries (mainly Finland), from other European countries, and from the Third World countries. The term immigrants is used in public politics, public debate, and public statistics, for all the four groups. Immigrants from the Third World countries, and their children, are also referred to as 'blackheads' ('svartskallar') – a word which has become notorious among ethnic minorities because it is often hurled as abuse at persons with dark hair. Since no statistics are available for specific ethnic groups, our paper is about the 1.6 million persons who in every day speech are called 'immigrants'. We shall, however, try to show differences in the situation for different ethnic groups wherever possible.

Local governance in the Swedish welfare state

Formally, local government in Sweden has a strong constitutional position, with extensive fiscal rights, a land-use planning monopoly, and extensive administrative responsibility with regard to the production and distribution of social welfare services, health, education, and technical infrastructure. In reality however, national government can and does exercise control over local governance, through legislation, regulations, and economic policy. Contradictory conclusions have been arrived at with regard to the scope of local governance (see, e.g., Strömberg and Westerståhl, 1984; Lane and Magnusson, 1987; Elander and Montin, 1990). Local governance is either facilitative or mandatory. In the first case, local governments make their own decisions. These include land-use planning, and some sectoral activities like culture and recreation. In the second case, legislation and statutory requirements regulate local governance. During the entire welfare growth period (1950-75), the mandatory sector, which includes social welfare, education, and health, expanded rapidly, and in the early 1990s accounted for about three quarters of total local government expenditure. However, municipalities exercise discretion by circumventing regulations (Khakee, 1983; Elander and Montin, 1990). Some policy areas, like foreign affairs, the labour market, and immigration and immigrant policy, are the sole responsibility of the national government. Municipalities are often required to execute specific tasks in these policy areas. For example, in the case of immigrant policy, the national government provides grants for local government to receive refugees and immigrants and to provide them with housing, education, and social services. Local government has often tended to do no more than what has been prescribed by the national government. For example, a recent public commission report says that only a tiny proportion of the refugees received by municipalities in the 1980s and early 1990s are self-supporting. Moreover, many of these refugees live in isolated ethnic ghettoes. Municipalities have failed to integrate them into the local community (Kuusela, 1993).

Another major aspect of local governance in Sweden has been the relative continuity and stability in policy making. Local governments have a similar political structure to the national government, with a municipal council elected every third year (from 1998, every fourth year). They have a central policy-making board, the 'executive board', which comprises representatives of nearly all political parties represented in the municipal council. In this sense the local system does not follow the national parliamentary pattern. The executive board has a coalition character (Elander and Montin, 1990; Reade, 1989). The policy-making boards have a number of subordinate departmental boards (that are similarly representative) to which the actual departments are responsible. Politics and administration remain close on account of the system of full-time councillors, many of whom also chair the major departmental boards. In big cities and towns there is an increasing practice of dividing full-time councillorship among major parties represented in the council. In this case, the coalition character is tempered by the influence of the parliamentary pattern (Gustafsson, 1993). This characteristic of local government has been detrimental for ethnic minority concerns. Ethnic minorities have so far not been able to organise themselves as a unified bloc. Furthermore, political parties have not made any real efforts to involve immigrants in politics. Their representation in municipal councils is very low in relation to their proportion of the population. In the 1994 elections, 93 first generation immigrants from non-European countries were elected to the municipal councils, i.e. 0.7 per cent of the total number of councillors, although this group formed 3.2 per cent of the electorate. The corresponding figure for the county councils was 7, i.e. 0.4 per cent of the total number of county councillors (Amin, 1996). Moreover, stability in local politics has meant reluctance either to take up 'blast effect' issues or to relegate them to administration.

A common description of Swedish politics once was the 'politics of compromise'. Swedes' general dislike of open political controversy, and assumption that problems could be solved objectively, were exemplified by the system of public commissions of enquiry. They produced reports on which legislation and public policy were usually based, and if they took time to reach their goals it was because their method of working involved a general practice of consulting all bodies officially acknowledged as entitled to speak for the specific interests affected. These were trade unions, employer alliances, tenant movements, and other so-called 'popular movements'. Almost the same procedure was repeated at the local level. The current development is towards 'dual democracy', where some issues involve the type of public discourse described above. In the case of an increasing number of policy issues, the duration of public enquiry is short, or proposals are presented by responsible ministers without any appreciable public consultation. This development implies that those groups which can look after their own interests, for example by lobbying, can influence various government organs. Ethnic minorities are at a disadvantage with regard to both types of policy process. Their representation in 'popular movements' is negligible. For example none of the three major trade unions has any immigrant on its board. The corresponding figure for the 50 largest companies is 1 per cent – often as foreign representatives (Amin, 1996). Since they are not organised as a political bloc and have no resources for lobbying, they are excluded from the decision-making process.

The 1960s and 1970s were characterised by a large range of regional policy measures seeking to enhance regional co-ordination and regional coherence between economically strong and weak regions in Sweden. The demarcation of stagnant areas requiring special attention, the creation of a regional system based on growth-pole ideas, and large scale transference of public money, were examples of this ideology of social cohesion and solidarity. The aim was that regions should have equal opportunities and uniform welfare programmes. Since the 1980s, this spatial ideology has given way to an ideology characterised by regional fragmentation and regional competition. EU-regionalism plays an important role in this context. Swedish regions are involved in 'space marketing' within Europe's new political geography. Ethnic minorities, like other socially disadvantaged groups, are regarded in this context more as a 'liability' rather than an 'asset'. There are no pronounced examples of ethnic discrimination in connection with urban regions' space marketing efforts. There are, however, indications within urban renewal programmes that increasing the attractiveness of a housing area may relate to a reduction in the number of less well-to-do families, including immigrant households.

The tenets of a national urban policy have been considerably weakened since the 1980s. Local governments have objected to national building standards and to statutory regulations for the provision of various services. Some controls, for example in the housing sector, have been dissolved. The Local Government Act has been amended twice in the last 15 years to provide greater freedom for the organisation of local governance and the management of public services (Fredlund, 1991). The state of local governance in Sweden in the 1990s reflects significant contradictions. While central regulations remain in many areas, and in some cases (for example provision of child-care and public libraries) new regulations have been or are being imposed, the national government, facing deficits in the national budget and the desire to join the European Monetary Union, is driven towards greater deregulation. The managed approach to development, with supporting social welfare strategies, is being gradually replaced by a proactive approach, where the state provides a framework within which firms, voluntary groups, and agencies deliver services. This interactive form of creating constituencies of support, and negotiating contributions, puts those who lack resources for organising themselves - among them ethnic minorities - at a considerable disadvantage. As a result, tensions and conflicts prevail, in urban regions which are increasingly being characterised by social polarisation, intercultural tensions, and social exclusion.

Growth of the ethnic minority population and racialisation of politics

Immigration to Sweden is of recent date. Moreover, immigrants have not arrived in a steady stream. Immigration levels have fluctuated because of domestic economic and social conditions and international political changes. In 1930, foreign-born persons numbered about 1 per cent of the population. Third World immigrants (from Africa, Asia and Latin America) numbered only 1,047 (Rojas, 1993). At the end of the Second World War some tens of thousands of refugees, from Germany and the Soviet-occupied Baltic states, migrated to Sweden. Most of them were quite easily assimilated into Swedish society. Between the 1950s and early 1970s, Sweden recruited foreign labour in order to satisfy the needs of the rapidly growing economy. Nordic immigration, especially Finnish, dominated this migration. The establishment of the Common Nordic labour market in 1954 made this easier. There was also a substantial influx of migrants from Italy, Greece, Turkey, and Yugoslavia. Manufacturing industries were short of labour, and a majority of this migrant labour easily obtained jobs, and local governments did not see any need for special provisions. The importing of labour reached a peak in 1970, when more than 75,000 persons migrated to Sweden. In the face of economic crisis and rising unemployment, steps were taken to stop non-Nordic immigration. These gradually led to a complete stop in this type of migration (Ålund and Schierup, 1991).

Immigration from Third World countries started with Chilean refugees in the early 1970s, followed by Syrians and Kurds from the mid-1970s, and Iranians and Ethiopians (mostly Eritreans) during the 1980s. European immigration consisted of Poles during the 1980s, and ex-Yugoslavians during the mid-1990s. A liberal refugee policy, granting asylum to so-called *de facto* refugees, and generous rules for family reunification, were the main reasons for this development (ibid., 1991). Immigration to Sweden, in relation to the country's population (8.45 million), has been relatively large. However, what have been conspicuous have been the government's decisions to grant resident permits to large groups of immigrants at the same time; for example, the decision to give asylum to 40,000 refugees from Bosnia in June 1993. Such decisions have aroused a major debate in the mass media, as well as in the Swedish Parliament, where immigration policy was seriously criticised.

In 1995, the number of foreign-born persons amounted to 936,022, or 11 per cent of the population; of these 640,616 were born outside the Nordic countries: 315,328 came from the Third World. Asian immigrants, mainly from the Middle East, number 215,059. The largest Asian groups are Iranians (51,629), Turks (35,948), Iraqis (25,105) and Lebanese (22,657). The black population is relatively modest and amounts to 39,768. The largest groups of black people are 14,251 Eritreans and 10,753 Somalis.

Immigrants are mostly concentrated in large cities. For example, 16.6 per cent of Stockholm County's population is foreign-born. In one of the municipalities of Stockholm County, Botkyrka, the foreign-born population is three times as high as the national average. In the age group 30-59 years, every fifth Stockholmer is foreign-born, and one child in four has foreign-born parents. Table 2.1 describes the number of foreign-born and 'blackheads' respectively in the ten largest towns in Sweden and the three metropolitan regions.

Immigrants, especially from Third World countries, are concentrated in a few Million Dwellings Programme housing districts (this state programme of housing development will be discussed later in the chapter). These districts have become ethnic ghettoes. In Stockholm, for example, these districts are Tensta, Rinkeby, Skärholmen and Fittja. Development of ethnic ghettoes has characterised the social exclusion process in Göteborg and Malmö as well as in other large Swedish towns. In Stockholm the number of Million Dwellings Programme areas with an ethnic concentration has increased from five in 1975 to nineteen in 1993

(Biterman, 1996). Other studies show that social segregation has increased during the 1970s and 1980s (see, e.g., Andersson & Molina, 1996). The concentration of immigrants in public housing has been quite heavy. Borgegård and Murdie (1996) show a strong relationship between the tenancy system and segregation. For example, very few immigrants from Third World countries - Chileans, Iranians, or Ethiopians - reside in owner-occupied or co-operative houses.

The Swedish immigration policy in the period since the Second World War has been described using three terms: assimilation, multiculturalism, and integration. During the 1950s and 1960s, immigrant policy was marked by the desire to assimilate the immigrants into Swedish society along the lines of the 'republican model'. According to this model the nation is a political community, and anyone willing to accept the rules of the polity, and adopt the national culture, can become a citizen (Castle and Miller, 1993). This policy coincided with other developments - rapid economic growth, labour scarcity, and extensive build-up of social welfare based on the universalistic principles of equality (Öberg, 1994). The labour immigration did not give rise to manifest problems. Most of the immigrants worked in manufacturing industry where no special requirements of Swedish language were necessary. There is very little research about how successful the assimilation policy was. In their study of Stockholm, Borgegård and Murdie (1996) show that this policy succeeded for groups with shorter cultural distance from the Swedes, e.g. Germans, Poles, and Italians. It was not successful with regard to labour immigrants from Greece, Turkey, or Yugoslavia, most of whom became increasingly segregated in the Million Dwellings Programme housing districts.

In 1975 the Swedish parliament voted for a new immigrant policy which to some extent concurs with the multicultural model, which, according to Castle and Miller (1993), accepts cultural difference and the formation of ethnic communities, and awards citizenship to anyone willing to accept the rules of the polity. In the Swedish version of the model, the creation of territorially distinct communities was not accepted. The Swedish policy emphasised freedom of choice with regard to the retention of minority cultural identity and partnership, calling for closer co-operation between immigrants and the native population (Government Bill 1975:26). This policy, however, was short-lived. Economic stagnation, increasing unemployment, and the development of a service economy, changed the premises for implementing the multicultural immigrant policy.

Industries no longer required newly arrived immigrant labour; the service sector required proficiency in the Swedish language, and 'social competence' of immigrants looking for employment; and the public sector no longer had the resources to let ethnic minorities retain their cultural identity, nor to build up the co-operation between Swedish popular movements and immigrants' associations.

With the increase of non-northern European immigrants, the character of the public debate about immigrant policy changed. Openly expressed racism, which was previously a taboo, became increasingly prevalent. Two events confirmed a widespread anti-immigrant feeling among the population. In 1988 the municipality of Sjöbo, in south-eastern Sweden, arranged a popular referendum about the central government's policy to persuade municipalities to receive refugee immigrants: 65 per cent of the electorate voted against it. In the public debate that followed the referendum, several municipalities voiced similar opinion: 'not on our doorstep'. This development put an end to the illusion of nationwide solidarity which was a prerequisite for the multicultural immigrant policy. Some municipalities questioned the policy of persuading local government to accept refugees in return for central government transferences. Even those municipalities which received refugees considered the whole proposal in economic terms. Refugee- and immigrant-reception created new jobs. There was hardly any discussion about the intrinsic value of receiving refugees, namely to help to create multicultural and vigorous local communities (Laukkannen and Östnäs, 1986).

The second incident was the election into the parliament of a racist party, Ny Demokrati, following the 1991 General Elections. Although Ny Demokrati failed to remain in parliament for more than one term of office, this was very much due to internal strife between the leading members of the party. Increasing numbers of opinion polls showed that public attitudes towards immigrants had changed since the 1980s. More and more Swedes were of the opinion that Sweden should receive fewer immigrants and refugees. The electorate of the leading political parties - the Social Democrats and the Moderates (the conservative party) - urged a more restrictive immigration policy (Demker, 1993; Demker and Gilljam, 1994).

Conscious of the significant racist vote, the non-Social Democrat Government between 1991-94, and the Social Democrat Government since 1994, have implemented nearly all the major policies put forward by Ny Demokrati. These include the forceful expulsion of immigrants who have for political or other reasons given false information about their actual identity, and an extremely restrictive immigration policy. The 1997 Immigration Act discards the use of the concept *de facto* refugees - the stretchable scope for humanitarian consideration. It also abandons the clause for granting asylum to military deserters, and the political-humanitarian proviso for granting asylum. Moreover, it sharpened the rules for family reunification (Government Bill 1997:25). Expulsions have included families which have lived in Sweden for several years. These measures have, on several occasions, been criticised both by UNHCR and the Humans Rights Watch during 1996 and 1997.

To some extent, national policy has been dictated by the spatial variations in the concentration of ethnic minorities. Their concentration in Stockholm, Göteborg and some other major towns in southern Sweden, and the endemic unemployment among immigrants, led to a rapid rise in the cost of social welfare. The national government has reduced transferences to refugee households waiting for resident permits. Moreover, suggestions have been made to reduce immigrants' rights with regard to special educational, social and cultural support (Guillou, 1996).

In the midst of implementing the most restrictive immigration policy since the Second World War, the Social Democratic Government in 1996 appointed a cabinet minister responsible for 'integration'. One of the experts appointed by the Integration Minister has defined integration as having an ambivalent position between assimilation and multiculturalism (Diaz, 1996). Another expert resigned from his post as adviser to the Minister of Integration, and accused the government of building up a sophisticated type of apartheid. He contended that, by cementing the division between immigrants and the Swedes, integration policy has failed in its objective to 'liberate' immigrants from their immigrant status (Dagens Nyheter, 1996). The Integration Minister himself seems to be quite uncertain about the concept of integration. He has said that he preferred the expression 'society of great variety'. This seems to be a meaningless semantic exercise, since the Minister has repudiated the idea of converting immigrants into Swedes (Dagens Nyheter, 1996). The most recent government proposal on immigrant policy rejects the concept 'multicultural society' in favour of 'society of manifoldness'. This simply means that the former creates expectations of a specific policy for ethnic groups, whereas the latter implies that ethnic issues can be solved by more general measures which applies to the entire population (SOU 1997:16). (Statens offentliga utredningar - SOU - means State Public Commission Report.) However, the government proposal does not discuss the long-term implications of this policy. A society of manifoldness implies that the society is more pluralistic, and that ethnic groups are not in any way hindered from finding life solutions of their own choice. This in turn requires a fundamental discourse about the future of the welfare society - something which the government is not keen to carry out.

Urban policy frames and their evolution in recent decades

Until the 1980s, Swedish urban development was characterised by the presence of a national urban policy. This consisted of a building code, together with statutory requirements relating to social welfare services, education, housing, and technical infrastructure. These were tied to state grants, which until recently were targeted at various policy areas, but which are now disbursed in the form of lump sums. The latter strategy marks a certain devolution of power (Fredlund, 1991).

The provision of social welfare services (child care, old-age care, health care) has been fairly comprehensive. Successive increases in the provision of these services followed the demands by different groups - households with children, the disabled, pensioners, etc. Ethnic minorities have not been vocal enough to have their special requirements met, except with regard to the provision of education in their mother tongue as well as Swedish. Despite these provisions, many immigrant youths have insufficient knowledge of the Swedish language. They go to schools located in ghettoes where they have few opportunities to intermingle with Swedish children. Since unemployment among immigrants, and especially among women, is very high, many immigrant households cannot send their children to day-care centres which in many ways contribute towards their later development (Holm, 1994). As regards old-age care, a recent survey by the National Board of Social Welfare showed that only 9 out of the country's 288 local governments have mentioned that immigrant pensioners require special services. Five of these nine municipalities only mention that such a problem exists. The other four have a longer survey of immigrant pensioners, but lack tangible proposals with regards to special services. Stereotyped conceptions seem to characterise local governments' considerations in this matter. Some municipalities reject the idea of what they consider 'differential treatment'. Others are of the opinion that in immigrant cultures, grown-up children look after their parents and relatives!

Health among immigrants is poorer than in the Swedish population. Many of the immigrants, even those with academic degrees, are compelled to take low status jobs, often with poor work environments and limited possibilities of affecting work conditions (SOU 1989:111). Moreover, many refugees have psychiatric problems, as a result of their internship in refugee camps, and their past experience of torture and political persecution. A recent survey shows that Yugoslavians rate their health conditions as six times poorer than the native Swedes do, while Chileans rate it five times, and Turks and Iranians three times, poorer (Socialstyrelsen, 1995). Another study shows that immigrants, especially so called 'blackheads', are more often ill than the Swedes. For example, Greeks and Yugoslavians have three times as many days with leave of absence owing to illness, Turks and Iraqis two and half times, and Chileans twice as many. Moreover, women in these groups tend to be ill more than men (SOU 1996:55, Aurelius, 1993).

Until the 1980s, the national government exercised extensive control with regard to the provision of housing. The major aims of the national housing policy were to eliminate housing shortage, keep housing costs at a reasonable level, discourage speculative building, and provide satisfactory housing standards. The Million Dwellings Programme, which took effect between 1965 and 1975, was a product of this policy. It was implemented with the help of municipal housing companies. But most of this housing was in the form of high rise apartment buildings, and was located in peripheral areas. The provision of essential services lagged behind the record level of housing construction. Most of the immigrants who arrived during and after this building boom were given apartments in the Million Dwellings Programme districts. Together with low income earners, these immigrants created socially and economically segregated areas. These suburbs are characterised by a large number of social problems: alienation, high turnover of tenants, boredom, vandalism, etc. As we shall describe below, several attempts have been made to renovate the Million Dwellings Programme areas, but most have had only limited success. In the meantime, ethnic concentration in some of these areas has accelerated, and they have attained the reputation of ethnic ghettoes. A recent study showed that youths living in some of these ethnic ghettoes, in Metropolitan Stockholm, feel that they are visiting a foreign country when they take the underground to the City (Rojas, 1995; Ålund, 1995).

The national urban policy, which enabled central government to direct the type, extent and location of development, was based on the premises of rapid economic growth and good public finance. In recent years the government has faced declining investments, soaring public debt, budget deficit, heavy unemployment, and all the other features of a stagnating economy. The national government's cutback strategies include the 'export' of the crisis of the welfare state to local governance. For example, the cost to local governments of social allowance has increased from 4.6 billion SEK in 1990 to 10.3 billion in 1994. The number of youths and immigrants who receive social allowance has increased most. These groups' dependence on social welfare benefits is going to increase more rapidly as the national government sharpens the rules for sickness and unemployment benefits (SOU 1996:34). Furthermore, the numbers living on social allowance have increased in large towns, especially in the three metropolitan areas (SOU 1989:67).

The ideological premises of the national urban policy were those of a provider state, committed to provide a comprehensive set of services to citizens regardless of their geographical location and ethnic background. Another important aspect of this policy was a belief in spatial coherence and regional co-operation. Since the mid-1970s these premises have changed. The shift from a provider state to a collaborative state implies that the state seeks new constituencies of support and new forms of partnership between state and market. The transposition from spatial coherence and co-operation to spatial fragmentation and competition implies that local governments are involved in promoting the development of 'nodes' and 'niches' (Healey et al., 1997). Ethnic minorities are neither regarded as suitable partners for national or local governance, nor as an 'asset' in place marketing.

Urban renewal: history and prospects

Urban renewal, involving comprehensive changes in buildings and land, has taken place in two separate periods. The first phase involved a large-scale renewal of town and city centres between the 1950s and the early 1970s, resulting in the demolition of residential buildings, the erection of business and shopping premises, and the widening of the existing streets - and construction of new ones - in order to increase accessibility for expanding car traffic. The second phase, which began in the 1980s, involves the renewal of suburban areas, especially the ones built during the Million Dwellings Programme period.

Hall (1979) presents a vivid picture of central city renewal in Stockholm:

> a block was vacated and cleared away, whereby the existing activities and business would be moved to previously developed blocks. Afterwards would come the widening of the streets, the possible building of a section of the tunnel and the drawing of cables. The many small plots of land would be combined into a few large units ... [and as a result] large parts of the CBD area acquired a wholly new physical structure, and in part a new economic, social and cultural content as well.

There was a strong regional bias in this large scale urban renewal. The implication was that such renewal was desirable in all the important regional cities or towns on an increasing scale: the larger the urban locality, the more important was urban renewal, with Stockholm's renewal being of importance for the entire country, since metropolitan Stockholm would then be able to compete with other European metropolitan regions. And in fact demolition in Stockholm exceeded that in any of the war-torn cities. This renewal led to a large-scale displacement of the population. The population of inner city Stockholm decreased from 440,000 to 236,000 between 1950 and 1985. The corresponding figures for Göteborg were from 266,000 to 142,000, and for Malmö from 144,000 to 94,000. The largest part of this displacement took place between 1950 and 1970 (Hall, 1991). Many of the less well-to-do households were compelled to move to the suburbs. For example, in Stockholm only every sixth household could obtain an apartment within the same parish (Schönbeck, 1994).

Important features of urban development in Sweden since the Second World War have been accelerated urbanisation in the whole country, and extensive migration from the so-called northern forest regions to the urban areas in southern Sweden - especially to the three metropolitan regions of Stockholm, Göteborg and Malmö. The urban population rose from 3.6 million (56 per cent of the total population) to 6.6 million (81 per cent of the total population). About one-third of the population came to reside in the three metropolitan areas. Urbanisation led to enormous construction of housing. Nearly 70 per cent of all housing in Sweden has been built since 1970. A large portion of this housing was built as a part of the Million Dwellings Programme (Vidén and Lundahl, 1992). As we mentioned above, it is in these Million Dwellings Programme housing areas that most immigrants came to be housed.

Urban renewal in the 1980s and 1990s has been concentrated in these areas. Although not more than 10 to 20 years old, many high-rise houses have become run-down because of the industrialised methods of construction, and previously untried building material, which have not stood the test of time. In many of the Million Dwellings Programme housing areas, the physical environment outside the buildings is very poor, with meagre space allotment and relatively poor provision of services. Many of these suburbs are reminiscent of the concrete slums built during the same period in the United Kingdom and elsewhere in Western Europe.

In the 1980s, urban renewal was initiated by national and local governments, co-operative housing companies, and other interests. The objective of this programme was the physical renewal of housing – improving the external physical environment, upgrading the standard of apartments, and completely rebuilding where required. One of the objectives was to attract more well-to-do households (Öresjö, 1996). This 'turn-around' policy was not successful from social, economic and ethnic perspectives. Development in areas with a high density of ethnic minority population shows that in Greater Stockholm, the proportion of immigrants in these renovated areas decreased. A majority of those displaced moved to areas where ethnic concentration was already very high.

The second phase of the renewal of the Million Dwellings Programme was initiated in the early 1990s, and is currently being implemented. There is a shift in emphasis in the renewal policy. The physical artefacts, in the form of buildings and external environment, are no longer the only, nor in fact the most important, elements in the renewal. The emphasis is more on the people who inhabit these areas, and the social and cultural relations prevailing among the inhabitants, and between the communities concerned and the world outside (Vidén, 1994). Social renewal implies measures to change and improve the housing areas' development potential, strengthen the political mobilisation, and reduce the crime and drug abuse as well as other social ailments (Jensfelt, 1993; Ehn, 1993).

Implications of urban renewal activity

Million Dwellings Programme areas around Sweden are of varying quality. Several of them had a bad reputation from the very start (Vidén and Lundahl, 1992). Some of these developed into socio-economic and ethnic ghettoes. Research about ethnic segregation has received significant interest (Biterman, 1996; Öresjö, 1996), and the national government has appointed public commissions about the discrimination of ethnic minorities in the labour market, and segregation in housing (SOU 1996:34; SOU 1996:55).

Research to date shows that there are both 'pull' (voluntary in nature) and 'push' (coercive in nature) factors affecting the development of segregated ghettoes. Pull factors include feelings of ethnic identity and kinship, possibilities of interacting with members of one's own ethnic group, and expectations about help to start a new life in a new society. Push factors include the discriminatory practices of landlords and housing companies, fear of being discriminated against in an area inhabited by native Swedes, and higher cost of living (Biterman, 1993). In many cases, immigrants are allocated housing in certain areas by a housing agency run by the local authorities. Swedish research shows that both pull and push factors prevail with regards to immigrants' residential decisions (Andersson-Brolin, 1984; Biterman, 1996).

Segregation implies that ethnic minorities' contacts with society at large are minimal. They lack the necessary social and economic networks. Entrance into the labour market is an important prerequisite for developing these networks. Allen and Hamnett (1991) demonstrate a close relationship between housing segregation and unemployment. Swedish Research shows that public labour market agencies play a minor role in the provision of jobs. About 70 per cent of jobs are arranged either though direct contacts with employers or through family and other social networks (Bel Habib, 1995). According to a recent study, unemployment among foreign-born is more than two-and a-half times greater than among native-born Swedes (Rojas et al., 1997). Requirement of being 'Swedish' have increased in the labour market. Formally, good knowledge of Swedish, and 'social competence', are given as major reasons for not employing immigrants. In reality, Swedish companies and even public agencies make many excuses for not employing persons with non-Swedish names. A recent survey of cases which were reported to the Discrimination Ombudsman showed that not even one case was considered to offend the public sense of justice (Alcala, 1996). The current legislation against discrimination in the labour market is not effective. The government is reluctant to introduce far-reaching legislation. Recently Michael Banton, the Chairman of the UN Committee against Racial Discrimination, suggested that Sweden should introduce registration of ethnic affiliation in order to counteract discrimination (a policy which has been increasingly employed in Great Britain). The Integration Minister dismissed the suggestion, by warning that employers should not be accused of being racists when they are in fact not so. Alcala (1996), in the above-mentioned study, rejects the present legislation by contending that it is based on the notion that the Swedes would get furious if immigrants were not moderately discriminated against!

Segregation has tended to accelerate in the past 10 to 15 years. Some Swedish researchers feel that development in Sweden will gradually resemble that in the United States and elsewhere, where segregated minorities live generation after generation in ghettoes and outside the mainstream. Social exclusion, or social marginalisation (Germani, 1980, uses the latter term), is not only a process but also a state of conditions. Germani uses three majors sets of factors to illustrate social exclusion: political, economic, and socio-cultural.

The implications of urban development activities on the social exclusion of ethnic minorities can be explored with the help of statistics regarding political, economic, and socio-cultural exclusion. In this section, we shall present some macro-statistics with regards to social exclusion of immigrants (Table 2.2). A more detailed discussion of the various aspects of this exclusion is put forward in the case study in the next section.

Except for membership of trades unions, which is not compulsory but is virtually automatic, the figures in Table 2.2 reveal significant political exclusion of foreign-born persons. Their electoral vote (i.e. turnout) is less than one-half that of the Swedes. Foreign-born persons are poorly represented on municipal councils – about half the numbers which might be expected if the composition of councils reflected the proportions of foreign-born persons in the population. The corresponding figures for county councils and parliament are 0.46 and 0.63. This is partly explained by the relatively low level of membership of political parties. The other important explanation is the length of residence in Sweden. A majority of the foreign-born persons elected to parliament, as well as those to the county and municipal councils, immigrated before 1968. This is not surprising, as it generally takes several years of endeavour before a person is elected to a political post. A recent study shows that it takes on average eight years for a person to progress from membership of a party to membership of a municipal board (Bäck, 1995).

Employment is extremely important with regard to the integration of immigrants in Swedish society. To obtain paid employment not only means the possibility of becoming self-supporting, but is also a speedy way to learn about the new country (Bel Habib, 1995). However, the picture with regard to employment is not very encouraging. Unemployment among the foreign-born is twice as high as for the entire population. In some groups, e.g. Somalis, nearly 80 per cent of those of working age are unemployed. Moreover, many immigrants are compelled to take menial jobs. These two factors explain the relatively greater number of poor households among the foreign-born as compared to Swedes. The same applies to the lower average income, which for foreign-born is about 65 per cent of that of Swedes. One of the major arguments by racist groups in Sweden is that immigrants cost a lot of money. Many of the refugees who arrived in the 1980s and 1990s have never had paid employment in the open labour market, which means that they are not entitled to unemployment benefits. This in turn increases their dependence on social allowance. Foreign-born households are nearly six times as likely as Swedes to be dependent on social allowance.

Putman et al. (1992) suggest that social participation is a prerequisite for fostering social capital. The latter simply implies mutual trust and understanding between people. These relations are built between people where they reside, work and spend their leisure time. Many foreign-born are members of various immigrant associations and clubs. However, these clubs are organised on ethnic grounds, often getting together people living in exile. They do not have the essential interplay with the Swedish popular movements. Immigrants' associations often have the function of retaining and reproducing their own specific ethnic culture (SOU 1996:55). Another drawback which immigrants experience in fostering social capital is the nature of their dwellings. Many immigrant households live in multi-dwelling housing, with little or no contact with Swedish neighbours. These conditions in turn reflect the higher degree of social isolation which foreign-born households experience. Many immigrant households end up in an downward spiral with no regular work, confined to ghetto dwelling and ethnic clubs. Health conditions deteriorate, and the opportunities to foster social capital diminish.

Rinkeby: Scandinavia's most multi-ethnic housing area

Situated about 10 km north-west of Stockholm City, Rinkeby, a sub-municipality within the municipality of Stockholm, has a population of 13,652 persons housed in 5,000 apartments, all of them high-rise multi-dwelling buildings. The area was built between 1969 and 1971 but the provision of services lagged behind. At the beginning of the 1970s the housing market was becoming increasingly segregated: with generous tax deduction possibilities, well-to-do households chose single dwelling houses, whereas poor households, which included many newly arrived immigrants, moved into Rinkeby and other Million Dwellings Programme Housing Areas (Sangregorio, 1989).

Of the population of 13,652 persons, 10,076 are foreign-born: i.e. 73.8 per cent of the population (compared with 18.7 per cent for Stockholm and 11 per cent for the whole of Sweden). Since 1980 there has been gradual change in the ethnic composition of the population: the number of immigrants from Nordic and other northern European countries has decreased, whereas immigrants from Asia and Africa have increased. The size of the Swedish population has decreased steadily, and in 1994 numbered 3,576, or 26.2 per cent of the total population. Table 2.3 illustrates the ethnic composition of Rinkeby.

The increasing concentration of immigrants in Rinkeby has resulted in a quarter of its tenants having a foreign background. The population is extremely heterogeneous, with over 100 languages spoken in the area. More than 90 per cent of the school children do not have Swedish as their mother tongue. The demographic composition is also different from that of other areas: children and youths make up 50 per cent of the population, and the percentage of people aged over 65 is lower than the national average. The average income in Rinkeby is the lowest, and the number of households on social welfare allowance the highest, in the whole of Stockholm. Unemployment is two and a half times higher than in the whole of Stockholm. In the last general election, electoral participation in Rinkeby was among the lowest in the whole of Stockholm County. Table 2.4 summarises some of the major variables with regard to social, political and economic exclusion in Rinkeby, compared with the whole of Stockholm.

Rinkeby's 5,000 apartments, all located in high rise buildings, are managed by three public companies: Familjebostäder, Svenska Bostäder, and Stockholmshem. In the middle of the 1980s, all three companies decided to renovate the housing stock: partly as a result of criticism from tenants, and partly because of the realisation that the Million Dwellings Programme buildings had poor construction. Whereas Svenska Bostäder and Stockholmshem decided on a relatively modest renewal programme,

Familjebostäder decided on an extensive renewal which would cost 650 million SEK and be phased over a period of five years.

According to the company the 'grey concrete buildings would be clothed in red bricks and apartments would be converted into maisonettes with bay-windows'. The company called all its tenants to a meeting, where the renewal plans were presented.

Despite the fact that there was no unequivocal criticism against the plan, those tenants who were present at the meeting were critical about the bay-windows, changes in the size of apartments, and 'brick clothing'. Moreover, many of the tenants did not like the idea of evacuating their apartments for a period of a year or more, nor the need to move twice – once to a temporary home, and then back to the renovated apartment. The major concern was the increase in the rents, which it was feared would be well beyond the reach of many households.

Familjebostäder, together with the other two public housing companies, notified a change in the tenancy system by introducing 'selective elimination'. The companies demanded that the local housing authority should allow only 'stable households, preferably Swedish ones. No more refugees or foreign born households with social problems should be given apartments in Rinkeby'. A heated discussion followed this proposal. Many immigrant households were worried that the new policy was aimed at reducing their number and increasing the percentage of Swedish households. Ehn (1993) suggests that Familjebostäder's use of the expression 'Rinkeby's new countenance' implied that the company wanted new faces, especially Swedish, in the renovated buildings. However, on the advice of the Ombudsman Against Ethnic Discrimination, the local authorities did not introduce such a regulation, although the tenants' association did support such discrimination!

The result of the Familjebostäder's renewal programme was a significant change in the composition of households. Nearly 65 per cent of the households affected by the renewal programme moved back to their previous apartments. Many of the relatively well-to-do Swedish and immigrant households decided not to move back, which resulted in an increased ethnic concentration. The company failed in its policy to attract Swedish households to move into the area, but succeeded in debarring households living on social allowance, by denying such households a lease. 70 out of the company's 1,245 apartments are currently vacant. But its policy has been copied in 23 other Stockholm housing districts, where households living on social allowance are no longer granted a lease of an apartment. Indirectly this policy has affected the housing situation of immigrant households, who to a relatively greater extent rely on social allowance. This has also resulted in foreign-born households increasingly relying on subletting (Andersson-Brolin, 1984).

Concluding reflections and guidelines for future research

As intercultural tension increases in many of Sweden's Million Dwellings Programme areas, the central and local governments find themselves trapped in a tight corner. On the one hand external factors, e.g. the harmonisation of social welfare policies within the European Union, and the market demands for a cut-back in the public sector, make every special provision for ethnic minorities unacceptable for the majority population. On the other hand, intercultural tension escalates as the situation for the immigrant population deteriorates. But this is by no means only a Swedish problem. With the shift from the 'provider' state to a neoliberal 'collaborative' state, this development exists in many Western European countries. The problem is exacerbated as regions and local places indulge in place marketing within the new political geography of Europe, as well as in the face of increasing competition from South East Asia.

Urban renewal and social exclusion are not only related to land-use and housing policies. The failure of the first round of urban renewal in the Million Dwellings Programme areas shows that these policies have to be accompanied by labour market, education, and other policies, if ethnic minorities are to have any chance of becoming integrated into Swedish society. The second round of urban renewal has a multi-policy approach. Its implementation is dependent on external as well as internal forces for change in society.

The Swedish developments in urban renewal and ethnic exclusion raise several topics for future research.

- Opportunities in the housing market are related to the cultural distance between various ethnic minorities and native Swedes. So far, very few studies have looked at the exact nature of cultural distance, and its impact on the social exclusion of different ethnic minorities. Public housing companies' social responsibility has declined considerably. More research is needed to examine how this development has affected the position of ethnic minorities with regard to a housing career.

- The Swedish political structure strongly favours the inception of general solutions which allow little scope for minority opinions. Further research is required to find out the implications of this factor, especially as the mental barriers between Swedes and immigrants increase. A recent survey shows that nearly half a million people are not 'found' in public statistics. They are not registered - either as 'unemployed', or as 'recipients of social allowance'. A majority of these are very likely immigrants. Research is needed in this respect, in order to find out whether uniform urban policies have forced these people to live outside the 'mainstream' of Swedish society.

- Research needs also to be focused on the exclusionary structure of local governance. Recent developments in urban regime theory show a different picture with regard to the division of power. The less-well-to-do, including immigrants, seem to be poorly represented in the coalitions and networks which control development of urban regions. Regime theory provides a useful theoretical framework for examining policies for the consolidation of multi-ethnic societies.

- Other, more specifically Swedish, issues which need to be researched further are the impact of the second round of urban renewal of the Million Dwellings Programme areas, central government's visions about the 'society of manifoldness', and the implications of social exclusion in the Swedish institutional context.

- Changes in the housing market depend on the shift from the provider state, based on universalistic principles of equality, to the collaborative state, with particularistic notions of welfare distribution. As Sweden and the rest of Western Europe develop new collaborative conditions, whereby the state looks for support from various coalitions and groups, social policies, including housing policy, are greatly affected. Comparative research is necessary with regard to the implications of this development for urban renewal and social exclusion.

Table 2.1 Foreign-born and black in ten largest towns

Towns/Regions	Total Population	Foreign-born	Black
Stockholm	711,119	120,746	59,513
Metro. Stockholm	1,725,756	286,569	130,952
Göteborg	449,189	78,408	43,510
Metro. Göteborg	770,375	105,110	53,870
Malmö	245,697	50,508	28,074
Metro. Malmö	817,022	109,661	54,144
Uppsala	183,472	22,811	13,324
Linköping	131,370	10,486	6,582
Norrköping	123,795	15,460	9,464
Västerås	123,728	17,631	7,194
Örebro	119,635	11,928	7,338
Jönköping	115,429	11,350	7,000
Helsingborg	114,339	15,600	7,333

Source: National Statistics Office of Sweden (1995), *Befolkningsstatistik 1995*, Statistics Sweden, Stockholm.

Table 2.2 Political, economic and social exclusion

	Swedes	Foreign-born[a]
Political Exclusion		
Electoral vote % (1994)	84	40
Representation on Municipal Councils (% of the electorate, 1994)	1.06	0.46
Membership of a Political Party (% of workforce, 1992)	12.2[b]	4.8
Membership of a Trade Union (% of workforce, 1992)	83.6[b]	80.5
Economic Exclusion		
Unemployment rate (% of workforce, 1994)	9.1[b]	17.6
Average household income (1994)	201,900 SEK	31,900 SEK[c]
% of families with income below subsistence level (1991)	8[b]	13.6
% of households receiving social allowance (1995)	5.7	23.9
Social Exclusion		
% of households owning own home (1990)	75.0	46.5
Average number of days with sickness benefit per employee (1990)	24.9	41.7
% of households which feel socially isolated (1991)	1.3	2.4
Membership of an association or club (% of population, 1992)	92.9[b]	86.5

Notes

a The Central Bureau of Statistics does not provide separate statistics for 'blackheads', which is unfortunate because it makes their problems less visible in the above statistics.

b Figures for the entire population including foreign-born persons.

c Figures for persons from Yugoslavia, Chile, Ethiopia, Iran, Iraq, Somalia, Turkey and Bosnia.

Source: National Statistics Office of Sweden (1996), 'Politiska resurser och activiteter 1978-1994', *Statistics Sweden*, Stockholm.

Table 2.3 Ethnic composition of Rinkeby (1994) (excluding Swedish-born)

Ethnic Minority	Total Size	Relative Share
Turks	2,097	15.4
Greeks	1,024	7.5
Finns	847	6.2
Iranians	697	5.1
Chileans	531	3.9
Ethiopians	460	3.3
Iraqis	393	2.9
Remaining (representing about 60 nationalities)	4,027	29.5

Source: National Statistics Office of Sweden (1995), *Befolkningsstatistik 1995*, Statistics Sweden, Stockholm (as Table 2.1).

Table 2.4 Some major characteristics of social exclusion in Rinkeby

Variables	Rinkeby	Stockholm
Population (1994)	13,652	703,627
Foreign-born population (1994)	10,076	131,397
% who were foreign-born (1994)	73.8	18.7
% of electorate voting in Municipal Elections (1994)	53.1	81.2
Average number of days with sickness benefit per employee (1994)	36.3	26.1
% of population who were gainfully employed (1993)	38.6	70.3
% of households receiving social allowance (1994)	40.6	10.2
% of Non-Nordic households receiving social allowance	68.0	37.0
Average household income (1993)	134,700 SEK	203,900 SEK
% of pensioners receiving early retirement pension % (1993)	10.5	6.4
% of households receiving public housing allowance (1994)	44.9	14.2
% of total workforce in blue-collar jobs (1990)	55.5	28.2
% of over-19 population with post-upper-secondary school education (1994)	10.0	33.0

Source: As Table 2.2.

References

Alcala, J. (1996), 'Lite diskriminering får man tåla', *Dagens Nyheter*, 10/8/1996:B3.

Allen, J. and Hamnett, C. (1991), *Housing and Labour Markets: Building the Connections*, Unwin Hyman, London.

Ålund, A. (1995), 'Ungdomar, gränser och nya rörelser' in *Rassmens varp och trasor. En antologi om främlingsfientlighet och rasism.* National Board of Immigration, Norrköping, pp. 54-75.

Ålund, A. and Schierup, C.-U. (1991), *Paradoxes of Multiculturalism. Essays on Swedish Society*, Avebury, Aldershot.

Amin, J. (1996), *Invandrarnas politiska representation*, Masters Thesis, Department of Political Science, Umeå University, Umeå.

Andersson, R. and Molina, I. (1996), 'Etnisk bostadssegregation i teori och praktik', in *SOU 1996:55 Vägar in i Sverige*, Allmänna förlaget, Stockholm.

Andersson-Brolin, L. (1984), *Etnisk bostadssegregation,* Swedish Council for Building Research, Stockholm.

Aurelius, G. (1993), 'Kulturell mångfald i vården – hot eller bot mot invanda rutiner' in Arnstberg, K.-O. (ed.), *Kultur, kultur och kultur – Perspektiv på kulturmöten i Sverige*, Liber Utbildning, Stockholm, pp. 179-191.

Bäck, H. (1995), *Att vara kommunalpolitiker på 90-talet,* Department of Political Science, Stockholm University, Stockholm.

Bel Habib, H. (1995), 'Att förstå främlingsfientlighet och rasism' in *Rasismens varp och trasor. En antologi om främlingsfientlighet och rasism.* National Board of Immigration, Norrköping, pp. 96-119.

Biterman, D. (1993), 'Invandrarnas boendesegregation', in Ehn, S. (ed.), *Så här bor vi – Om invandrares liv och boende,* Swedish Council for Building Research, Stockholm, pp. 21-40.

Biterman, D. (1996), 'Den etniska boendesegregationens utveckling i Stockholms län 1970 – 1990', in Bohm, K. and Khakee, A. (eds), *Etnicitet, segregation och kommunal planering,* Nordplan, Stockholm, pp. 40-63.

Bohm, K. and Khakee, A. (eds) (1996), *Etnicitet, segregation och kommunal planering*, Nordplan, Stockholm.

Borgegård, L-E. and Murdie, R.A. (1996), *Immigration, Residential Segregation and Housing Segmentation in Metropolitan Stockholm 1960-95*, Research Report for the ENHR Conference, August 26-31, 1996, Helsingör, Denmark.

Castle, S. and Miller, M. (1993), *The Age of Migration*, Macmillan, London.

Dagens Nyheter (1996), 'Sverige anklagas för apartheid', *Dagens Nyheter,* 9/12/1996:A11.

Demker, M. (1993), 'Stäng gränserna? Svenskarnas åsikter om flyktingmottagning' in Holmberg, S. and Weibull, L. (eds), *Perspektiv på krisen, SOM-undersökning 1992:9*, Department of Political Science, Göteborg University, Göteborg.

Demker, M. and Gilljam, M. (1994), 'Om rädslan för det främmande', in Holmberg, S. and Weibull, L. (eds) *Vägval, SOM-undersökning 1993:11*, Department of Political Science, Göteborg University, Göteborg.

Diaz, J.A. (1996), 'Invandrarnas integration - några teoretiska och metodologiska utgångpunkter' in *SOU 1996:55 Vägar in i Sverige*, Allmänna förlaget, Stockholm.

Ehn, S. (1993), *Så här bor vi. Om invandrares liv och boende*, Swedish Council for Building Research.

Elander, I. and Montin, S. (1990), 'Decentralization and control: central-local government relations in Sweden', *Policy and Politics,* vol. 18, no. 3, pp. 165-180.

Fredlund, A. (ed.) (1991), *Swedish Planning in Times of Transition*, The Swedish Society for Town and Country Planning, Gävle.

Germani, G. (1980), *Marginality*, Transaction Inc., New Brunswick, N.J.

Goldfield, D.R. (1979), 'Suburban development in Stockholm and the United States. A comparison of form and function', in Hammarström, I. and Hall, T. (eds), *Growth and Transformation of the Modern City*, Swedish Council for Building Research, Stockholm.

Government Bill 1997:25 *Svensk migrationspolitik i globalt perspektiv*, The Swedish Parliament, Stockholm.

Government Bill 1975:26 *Riktlinjer för invandrar- och minoritetspolitiken*, The Swedish Parliament, Stockholm.

Guillou, J. (1996), *Svenskarna, invandrarna och svartskallarna,* Pan Norstedts, Stockholm.

Gustafsson, G. (ed.) (1993), *Demokrati i förändring*, Publica, Stockholm.

Hall, T. (1979), 'The central business district: Planning in Stockholm', in Hammarström, I. and Hall, T. (eds), *Growth and Transformation of the Modern City*, Swedish Council for Building Research, Stockholm, pp. 181-232.

Hall, T. (ed.) (1991), *Planning and Urban Growth in the Nordic Countries*, E. & F.N. Spon, London.

Hammar, T. (ed.) (1985), *European Immigration Policy. A Comparative Study*, Cambridge University Press, Cambridge.

Healey, P., Khakee, A., Motte, A., and Needham, B. (1997), *Making Strategic Spatial Plans. Innovation in Europe*, UCL Press, London.

Holm, T. (1994), *Invandrarna i den kommunala planeringen*, Masters Thesis, Department of Political Science, Umeå University, Umeå.

Jensfelt, C. (1993), 'Den etniska mångfaldens Fittja. Reflektioner kring ett kommunalt förankrat förbättringsarbete' in Ehn, S. *op. cit.*, pp. 41-56.

Khakee, A. (1983), *Municipal Planning. Restrictions, Methods and Organizational Problems*, Swedish Council for Building Research, Stockholm.

Kuusela, K. (1993), *Integration i invandrartäta bostadsområden?* Department of Sociology, Göteborg University, Göteborg.

Lane, J-E. and Magnusson, T. (1987), 'Sweden' in Page, E.C. and Goldsmith, M. (eds), *Central and Local Government Relations: A Comparative Analysis of Western European Unitary States*, Sage, London.

Laukkanen, M. and Östnäs, A. (1986), *Invandrarlandet Sverige – det nya folkhemmet?* Studentlitteratur, Lund.

Öberg, N. (1994), *Gränslös rättvisa eller rättvisa inom gränser? Om moraliska dilemman i välfärdsstatens invandrings- och invandrarpolitik*, Department of Political Science, Uppsala University, Uppsala.

Öresjö, E. (1996), *Att vända utvecklingen – Kommenterad genomgång av aktuell forskning om segregation i boendet*, SABO Utveckling, Stockholm.

Putman, R.D., Leonardi, R. and Nanetti, R. (1992), *Making Democracy Work: Civic Traditions in Modern Italy*, Princeton University Press, Princeton.

Reade, E. (1989), *Britain and Sweden: Current Issues in Local Government*, Almqvist & Wiksell, Stockholm.

Rojas, M. (1993), *I ensamhetens labyrint. Invandring och svensk identitet*, Brombergs, Stockholm.

Rojas, M. (1995), *Sveriges oälskade barn – Att vara svensk men ändå inte.* Brombergs, Stockholm.

Rojas, M., Carlsson, B. and Bevelander, P. (1997), *I krusbärslandets storstäder. Om invandrare i Stockholm, Göteborg och Malmö*, SNS, Stockholm.

Sangregorio, I-L. (1989), *Rinkeby: Bygg inte bort livet*, Boverket, Karlskrona.

Schönbeck, B. (1994), *Stad i förvandling. Uppbyggnadsepoker och rivningar i svenska städer från industrialismens början till idag*, Swedish Council for Building Research, Stockholm.

Socialstyrelsen (1995), *SOS-rapport*, National Board of Social Welfare, Stockholm.

SOU 1989:67 *Levnadsvillkor i storstadsregioner*, Allmänna Förlaget, Stockholm.

SOU 1989:111 *Invandrare i storstad,* Underlagsrapport från storstadsutredningen, Allmänna förlaget, Stockholm.

SOU 1996:34 *Aktiv arbetsmarknad*, Allmänna Förlaget, Stockholm.

SOU 1996:55 *Sverige, framtiden och mångfalden*, Allmänna Förlaget, Stockholm.

SOU 1997:16 *Sverige, framtiden och mångfalden - från invandrarpolitik till integrationspolitik*, Allmänna Förlaget, Stockholm.

Strömberg, T. and Elander, I. (1993), 'Från citysanering till den måttfulla staden', *Nordisk Arkitekturforskning*, vol. 6, no. 1, pp. 99-113.

Strömberg, L. and Westerståhl, J. (1984), *The New Swedish Communes. A Summary of Local Government Research*, Department of Political Science, Göteborg University, Göteborg.

Vidén, S. and Lundahl, G. (1992), *Miljonprogrammets bostäder. Bevara-Förnya-Förbättra*, Swedish Council for Building Research, Stockholm.

Vidén, S. (1994), *Stadsförnyelse och bostadsombyggnad – Att söka kunskap för varsam förbättring*, Department of Architecture and Urban Development, Royal Technical College, Stockholm.

3 Living on the margins of society:
Ethnic minorities in the Netherlands

TON VAN DER PENNEN, VINCENT SMIT AND
JAN WILLEM VAN DE WARDT

Ethnic minorities in the Netherlands

Origins

The minority groups that now demand the greatest policy attention in the Netherlands have settled there only recently. Three waves of migration can be distinguished. The first started in the 1960s and 1970s and was composed mainly of Turkish and Moroccan men who came to work in the Netherlands. They settled around industrial centres, and at first everyone thought their stay would be temporary. Large proportions of these pioneer groups did indeed return home after their contracts expired. From the mid-1960s onwards, however, re-migration slowed, and a process of family reunification, or, put more generally, a social migration process, got underway. The initial settlement patterns remained unaltered, and as of 1996 most ethnic minority groups still live in the cities of Amsterdam, The Hague, Rotterdam, and Utrecht.

There were different underlying reasons for the second wave of migration, made up of Surinamese and Antilleans. The Dutch cities had long been a focus for individual students, but the nature of the migration stream changed from the 1960s onwards, when the economic prospects of these colonies began to dim. Notably, a large wave of immigration occurred on the eve of Surinamese independence in 1975. In this case, too, the stream of immigrants followed the old patterns, settling in the course of time in the big cities, despite a nationwide dispersal policy which required local authorities to provide housing for the newcomers.

The third stream of migration is more diffuse, involving a disparate collection of asylum-seekers and refugees, most of whom arrived in the early 1990s. Though many failed to secure official residence permits, a sizeable group had better fortunes and were admitted on a semi-permanent basis. An unknown number of others chose to remain in the country illegally. These two groups are not included in the analysis to follow, because little research material is available on the position they occupy in Dutch society.

On the basis of ethnic origin, 15.6 per cent of the 1992 population of the Netherlands were not ethnic Dutch. 6.1 per cent of the total population in that year was from the more distinct ethnic minority groups, up from 5.5 per cent in 1990. The four largest groups – Surinamese, Turks, Moroccans, and Arubans/Antilleans – together accounted for over 85 per cent of the distinct ethnic minority population in 1992.

Patterns of settlement

The minority groups most targeted by policy – Surinamese and Antilleans, Turks, and Moroccans – reside predominantly in the four largest cities in the country: Amsterdam, Rotterdam, The Hague, and Utrecht (see Table 3.1). In these cities they show a tendency towards concentration. Indices of segregation at neighbourhood levels range from 24 in Utrecht to 51 in The Hague (CBS - Centraal Bureau voor de Statistiek - municipal population statistics, derived from Tesser et al., 1996, p. 64). No major shifts have occurred in the past fifteen years. The segregation of Turks and Moroccans is stronger than that of other groups, and a relatively sharp segregation also exists between them and Surinamese and Antilleans. Tesser and colleagues (1996, p. 88ff) have made an overall comparison of inner-city segregation in Brussels, London, Manchester, Frankfurt am Main, and Düsseldorf. They concluded that the indices of segregation for Dutch cities are comparable to those in the two English cities; they are lower than those for Brussels, and higher than those in the two German cities.

The social position of migrants

Spatial mismatch

Migrants occupy an unfavourable position in Dutch society. They suffer a high degree of unemployment, and they live in the poorer sections of the housing stock. They are subject to social exclusion. In the political debates on this issue, people warn of the danger of a social divide. Ethnic segregation and concentration are considered undesirable conditions.

One type of explanation of exclusion focuses on the characteristics of ethnic minority people: especially their levels of income and education. Another draws attention to the milieux they live in, which are said to generate resignation and apathy – a culture of poverty (Lewis, 1966). Other studies, e.g. Wilson (1987), have pointed to the influence of residential areas, notably ghettoes. All such explanations concentrate on the behaviour people develop which leads to exclusion. While such explanations may provide worthwhile insights into things like crime and social safety in particular neighbourhoods, they are also the subject of some controversy, and have been accused of blaming the victim. We will not deal with these approaches here. Our approach is in line with the spatial mismatch theory formulated by Kassarda (1985). It holds that a number of demographic and economic trends serve to force certain population categories into the margins of the urban community. This perspective puts heavy emphasis on exclusion in the labour market. In the Netherlands this especially affects migrants. High unemployment, a key cause of the social exclusion of migrants, can in large part be accounted for by spatial mismatch. The assumption is that rapid deindustrialisation of urban economies has substantially decreased the number of jobs for which minorities are qualified. This 'exclusion' can be exacerbated by the selective migration to the cities of those minority people who have low levels of schooling.

Demographic trends A process of suburbanisation has been underway in the Netherlands since World War II. It is a selective migration, made up predominantly of higher-income families with children, who trade the city for a suburban environment. It coincided with a process of selective immigration into the cities: the advent of unskilled immigrants. However, their chances of earning an income simultaneously diminished in the rapidly changing urban economy. As a result of the two streams of migration, the proportion of households leading an unemployed existence has sharply risen in Dutch cities.

Economic trends There has been a general decline in the manufacturing sector, as well as a de-concentration of manufacturing activities away from the cities. The general decline is a result of overall changes in the nature of Western economies, where the importance of manufacturing industry has diminished in relation to the service sector. At the same time, the share of urban manufacturing in the Netherlands has decreased in relation to the total manufacturing sector. This is a consequence of the de-concentration of economic activities, which in turn is partly a function of better transport facilities existing or being created elsewhere. In the 1970s, this led to a widespread loss of jobs in the four large Dutch cities; in the 1980s the service sector grew there by leaps and bounds.

The nature of unemployment

Figure 3.1 plainly shows that unemployment is far higher in the four largest minority groups than among the ethnic Dutch population. This is true of both men and women, and was the case in the entire period from 1987 to 1994. Among the ethnic Dutch, as well as among immigrants from EU countries and the USA, female unemployment is substantially higher than male unemployment, but in the four large ethnic minority groups, unemployment affects women and men in equal measure.

Over the entire period depicted, unemployment describes a weak U-shaped curve, as is clearest for the ethnic Dutch population. Their unemployment gradually eased until 1991, and then climbed back to its 1987 level. Careful attention discloses a similar development among the ethnic minorities.

The low level of education prevailing among the potential ethnic minority labour force is part of the explanation for their considerably higher unemployment rate. Even among people with comparable levels of schooling, however, the risk of unemployment is far greater for ethnic minority people than for the ethnic Dutch (Niesink and Veenman, 1990; Kloek, 1992). Many observers have drawn the conclusion that ethnic minority people face disadvantage and discrimination in the labour market (Veenman, 1991). They argue that discrimination occurs in the recruitment, selection, and sacking of personnel. Although discrimination in such situations is typically hard to prove, there are indications that employers do have a preference for ethnic Dutch employees.

At the beginning of the 1980s, unemployment in the cities did not deviate much from national figures; but between 1982 and 1984 it climbed to way above the national average. Breaking down the unemployed by educational level, it can be seen that for all educational categories, rates in the cities are higher than they are nationwide. The scale of unemployment at the lower educational levels in the four cities is associated in part with the numbers of jobless people from ethnic minorities.

Figure 3.2 shows how migrants are over-represented at the bottom of the labour market. We might call this an ethnification of the lower ranks of the labour market. Turks and Moroccans in particular have fallen victim to the rapid changes in city economies. The greatest demand for labour is to be found in the expanding sectors in the service industries. Given that migrants are oriented to the manufacturing sector, this provides us with one major explanation for their joblessness, and hence their exclusion from the metropolitan communities.

Yet the recorded growth figures cannot be explained solely by dwindling job opportunities in the manufacturing sector, because that loss of jobs was largely offset by growth in the service sector. The explanation has to be sought primarily on the supply side of the metropolitan labour markets. This applies not only to Dutch cities, but equally to other European countries and the USA (Kassarda et al., 1992; Kloosterman, 1994). Unemployment problems are to be found predominantly at the bottom end of the labour market. The notion of the bottom end of the labour market can refer to human characteristics as well as to job characteristics (SCP 1995, pp. 139-141). The data in this section refer to human characteristics. Of these, level of schooling can serve us as the foremost predictor of whether someone will be near the bottom of the labour market. The information in Figure 3.2 makes clear that unemployment in the cities is largely, though not exclusively, the problem of people whose educational level puts them at the bottom of the market. The high unemployment at the bottom end of the metropolitan labour markets is a key component in the thesis of spatial mismatch. As we have seen, the assumption is that a significant proportion of the unskilled jobs on which people at the bottom end of the market depend have vanished with the rapid deindustrialisation of the city economies. But the high unemployment in the low ranges of the labour market is influenced by at least two additional factors. It can increase as a result of selective immigration to the cities by people with relatively little education who are mainly from ethnic minorities, as this augments the supply of labour at the bottom end of the market. But unemployment can shrink if new jobs are created in non-manufacturing sectors for which people at the bottom of the labour force are qualified.

In addition to specifically targeted labour market policies, the various levels of government also try to address the spatial mismatch through urban renewal.

Urban renewal as a strategy against urban decline

At present, urban renewal is seen as a major policy strategy for tackling the problems just outlined. We will now have a closer look at this strategy, first introducing the actors and then describing the policies.

The policy network of urban renewal

In this section we briefly examine the roles of the various parties involved in urban renewal, urban revitalisation, and housing rehabilitation, in the Netherlands. The most important ones are national government, local authorities, housing associations, private landlords, property developers, and community residents and their organisations. Before we can clearly set out the roles of these parties, and the shifting relationships between them, a few remarks are in order about the position of the public housing sector.

In the past fifty years, the institutional framework in which the expansion, renewal and rehabilitation of Dutch cities has occurred has been heavily dominated by government. This hegemony of the public sector, which arose after the Second World War, was brought about by a regulated planning system, and extensive public funding of the non-profit rental sector. A turning-point was reached in the late 1980s, when spending cuts, and disillusionment with the outcomes of neighbourhood housing rehabilitation, led to a revaluation of private initiative. More leeway was created for property developers, the construction of business complexes was stimulated, housing associations were given the freedom to act independently of public authorities, and home ownership was increasingly encouraged. Given such developments, it will be clear that relations between the various actors on the urban renewal scene are not static, but shift over time. Let us have a brief look at these actors.

The national government still plays a particularly vital role, albeit at some distance. Its decisions largely determine the context in which others can act, and the sheer affordability of certain policy options. Through its financial inputs into the rental sector, the housing rehabilitation in older city districts, and the metropolitan infrastructure, the government has exerted enormous influence over the existing spatial situation. It additionally sets regulations, furnishes planning instruments, and stimulates or blocks the building of new towns and development areas. Since the late 1980s the national government has significantly reduced its direct intervention as far as funding and regulation is concerned. Nonetheless, through its stringent planning policies it still wields great influence, largely controlling the allocation of large building sites outside existing urban areas.

Local authorities play an important part in the implementation of national government policies, and they influence the planning and management of urban space in other ways too. Below, we will examine shifts that have occurred in such powers. Suffice it to say here that local authorities have become increasingly dependent on other parties in fulfilling their tasks. Since the 1970s they have had to take increasing account of the standpoints of residents, and in the 1990s they must also gain the support of housing associations and various other parties, including property developers.

Housing associations control a vast section of the housing stock in the large Dutch cities – in Amsterdam 57 per cent of the total. Their policy latitude has been greatly expanded since the early 1990s: they have increasing freedom to operate autonomously, especially now that direct central government funding has been ended in return for the purging of their debts. Their policies are crucial to the position of minorities, who overwhelmingly depend on them for housing in the subsidised rental sector.

The role of private landlords continues to diminish. The nationwide rent policies of the 1950s and 1960s made it difficult for them to break even, and many withdrew. In the 1970s they sold large portions of their rental stock, much of it in bad repair, to owner-occupiers, or to local authorities during rehabilitation projects.

Property developers have been acquiring an important position in recent years. Their projects directed at trade and industry are increasingly valued by local administrators, who are concerned about high unemployment and the financial base of the municipal budget. The role of developers in housing construction is also growing, especially since local authorities are setting greater store by the creation of housing for higher-income groups – the realisation of which they leave to property developers.

Community residents and their organisations have also played a major role, especially in the 1970s and 1980s. This arose from resistance to plans for city core development in the 1960s, and from demands for a say in large-scale rehabilitation projects in the 1970s. Direct resident involvement now appears to have subsided, but their professionalised, subsidised organisations have meanwhile acquired a place in metropolitan policy networks. Residents basically have the following possibilities to influence policy. They can take part, either directly or through their organisations, in a range of participation procedures. They can lodge objections against plans. They can start action committees or appeal to community organisations. A more formal route of influence is through their rights to elect city and district councils. Of importance here is that, for about ten years now, residents of non-Dutch nationality have been eligible to vote and stand for election at the local level. No research has come to our attention yet on the actual policy influence that minority groups may have gained from this, but our knowledge of city policy networks, and the low interest among minorities in local elections, suggest that such influence is still extremely limited. The same applies when we look at the influence of ethnic minority organisations, or at the influence minorities have within traditional community organisations (see Case Study 1). And such is even more starkly the case with regard to their influence on urban revitalisation policies.

Urban policy frames

The parties reviewed in the previous section operate in the metropolitan policy framework. We shall now go on to describe, in an historical perspective, the policy ideals that were dominant in three different periods, and the implications such thinking had for city residents, and for migrants in particular.

Urban renewal in the Netherlands has assumed three distinct forms since World War II. In the period 1945-74, emphasis was on the reconstruction and economic stimulation of the city. It was a period of intensive city core development. In the second period (1974-89), concern for urban economic development shifted to the older urban districts, with the creation of Building for the Neighbourhood programmes. Since 1990, attention has turned to revitalisation of the city, with emphasis on both economic and social aspects of urban development.

City core development In the decades following the Second World War, policy on housing in city centres was ruled by the belief that living there was attractive only to a tiny cultural elite. Urban expansion was providing housing to ever larger groups of city dwellers. Essentially such plans were in line with the CIAM conception (Congrès International d'Architecture Moderne) (see Case Study 2), which made sharp distinctions between the functions of residence, recreation, and work. Work was to remain in the city and its immediate environs, and large-scale projects were implemented to stimulate it. The older urban districts, containing mainly low-cost housing, were targeted for demolition, and for replacement by large-scale office construction, shopping precincts, traffic arteries, and other such facilities. In addition to the reconstruction of urban areas, and the redesignation of existing residential areas for economic core functions, there were also clearance proposals for urban residential areas. They included the building of new housing, and a drastic revamping of the existing urban structure.

Housing and recreation were developed mainly at the edge of town, or in housing estates. This urban expansion consisted of high-rise buildings and single-family dwellings, and resulted in a massive exodus from older urban districts, especially by families with children who opted for family dwellings 'out amongst the greenery'. It was a selective migration of families, namely of more affluent households (Smit, 1991). According to the then prevailing viewpoint, the people who remained, who had moved into the older districts, were people who could not afford an alternative. Such districts were referred to as deprived neighbourhoods, where poverty and lack of opportunity were concentrated, and they are the neighbourhoods where the first immigrants from Mediterranean countries came to live (see earlier discussion in this chapter). These demographic developments could proceed due to a transformation in urban thinking.

'Building for the Neighbourhood' A turning point in the thinking about urban development was reached in the 1970s. While city core development had received priority until then, emphasis now shifted to the strengthening of the residential function of the old inner-city quarters.

This can be identified as a primary tenet of the ascendant policy vision. It concerned the preservation, restoration, and reinforcement, of the existing residential environment in the cities. Strengthening the residential function, rather than promoting the economic function, came to be a central premise of urban development thinking. The cities now entered a second phase of reconstruction, directed at the social aspects of the residential environment. Since a large proportion of those living in run-down neighbourhoods were thought to have no other choice than to go on living there, a detailed, intensive, and integrated plan of action had to be implemented. Practicability was considered more important than large-scale spatial conceptions. The depopulation and impoverishment of the cities was defined as a key problem, and it was partly the outcome of clearances, and of the construction of infrastructural facilities and office blocks. Subsidised housing - building for the financially weak - was now paramount. The suburbanisation of the more affluent segments of society was not encouraged, but it was seen as an inescapable part of the solution to the quantitative housing shortage in the cities.

The second policy tenet was the preservation of the existing city structure: the notion of the compact city. This contrasted with urban reconstruction programmes of the preceding period, which were based on separation of functions and on low-density development. The functionalist principles that had governed planning were abandoned, and the compact city was to be shaped by a blending of functions and by high-density development. Comparatively little attention was paid to the unsnarling of bottlenecks in the vehicular accessibility of inner cities, and there was a taboo against urban core development. Earlier traffic artery projects already in implementation were critically reassessed, and many were halted or cancelled.

An increase in the scale of shopping facilities continued throughout the period, but the sustaining of small-scale facilities at the neighbourhood level was the guiding principle. Atmosphere and sociability, rather than variety of shopping facilities, seemed of overriding importance.

The prevailing policy vision was also an outcome of community activism. In the social climate of democratisation and emancipation, residents of old city quarters organised into action committees and neighbourhood development bodies, to resist the ravaging of their homes by 'citification'. They became important negotiating partners for the local authorities in urban renewal. The third major policy tenet (we have already mentioned the residential function and the 'compact city') concerned its social and emancipatory goals. This was likewise in contrast to the previous period, when residents had been treated as consumers.

Urban policy targeted, first of all, the 'financially weaker citizens' in the older city districts, or it tried to give preference to those whose freedom of choice on the housing market was the most limited. Secondly, it endeavoured to preserve the existing social contexts in the older neighbourhoods, giving preferential treatment to those already present. Thirdly, greater political power in local decision-making processes was promoted for these 'disadvantaged' sections of the community.

Spatial, 'technical' interventions were seen as means and preconditions for community development, and for easing deprivation in a whole range of social domains. Although the apparatus was geared to 'building for the neighbourhood', ancillary measures were taken to ameliorate the deprived condition of the existing residents.

To summarise, the underlying assumption in this policy era was that deterioration of old city neighbourhoods was caused by two mutually reinforcing processes: a material process of neglected maintenance, and a social process of alienation, disintegration, and impoverishment. Urban restructuring was therefore to include both renovation and innovation. Its objective was to upgrade the housing stock, recognised as vital to the residents, to a higher level of quality, as well as to furnish those residents with better opportunities for social integration urban renewal as a campaign against social deprivation. This resulted in the notion of Building for the Neighbourhood, a logical outcome of earlier community resistance to plans for urban clearance in neighbourhoods, and to what was called by its opponents 'deportation to outlying districts and dormitory towns'.

Exclusion? Because immigrants developed into a major category of city residents in that same period, one might infer that social inclusion occurred. In part this was indeed the case. Looking back from our current vantage point, we can argue that the plans developed on the basis of the compact-city concept improved the housing situation of migrants. When they settled in the Netherlands their housing situation was far inferior to that of long-time residents, and the gap has since narrowed substantially (Tesser et al., 1996). Their housing conditions nevertheless still lag behind those of the ethnic Dutch.

Yet there is also evidence of social exclusion. In the early stages of Building for the Neighbourhood, the migrants' housing situation was poor:

they lived in the worst parts of the neighbourhoods. Ethnic Dutch residents came to regard this, and the subsequent influx of migrants (partly for family reunification) as a sign of neighbourhood decline. Immigrants were not yet being targeted by the Building for the Neighbourhood policies designed to upgrade housing conditions. Such policies applied only to official dwellings, and immigrants lived mostly in rented rooms. Few of them held tenancy even in poor-quality official dwellings, and they were therefore in no position to make demands for renovation, or for rehousing to newly built dwellings with affordable rents.

Their Dutch neighbours certainly did do so, and with considerable success. At this point the mechanism of the unequal negotiating position of the ethnic minorities began to operate. They were not residents of official dwellings, and they were not party to the political lobbying necessary to achieve favourable decisions on housing and neighbourhood renewal; they could merely observe what was happening, and see what housing would be left over. The ethnic Dutch, for their part, got on the gravy train of urban renewal, and a broad coalition between neighbourhood residents, local authorities, and housing corporations, created large-scale projects to thoroughly refurbish old buildings, or to replace them with new, subsidised housing. To the local authorities, the co-operation of residents was essential for the operation to succeed; in return for that co-operation the ethnic Dutch residents demanded affordable rents and a housing allocation procedure tailored to their interests. Although the 'impartial' rules of allocation that were formulated contained no openly discriminatory elements, everyone involved was aware that a requirement like 'a minimum of ten years' residence in the district' worked to the advantage of ethnic Dutch residents. Many of them were also for such measures because they saw urban renewal as a means to give the neighbourhood a more strongly 'Dutch' character. Paradoxically, however, a side-effect of such rules was that while many old, steadfast residents took their places in the high-quality housing, the older housing stock they vacated emerged as a sudden, unexpected reservoir for groups who did not qualify for the 'real' business of urban renewal: young people, and ethnic minorities.

In the first ten years of urban renewal, up to the mid-1980s, minorities were a group consigned to the rear of the train. But things were to change. Having appeared in the housing market in the 1970s, ethnic minorities had now become a category of households with tenants' rights. Though residing in old, cramped, low-demand flats, they could now begin navigating the market as priority-need candidates, or tenants with rights to rehousing.

Their length of residence and official tenancy in council housing, and subsequently also in housing corporation flats, had strengthened their bargaining position. And after ten years of urban renewal, the belief that urban renewal measures could also bend the demographic composition of areas towards increasing Dutch ethnicity had been shaken. It was clear that was not how it worked. Pressure was now building up from ethnic minorities, researchers, and various organisations, for genuinely equal treatment for minorities, and discrimination in the housing market became a topical issue in politics and public opinion. Open expressions of preferential treatment for people of Dutch origin were publicly censured. The housing needs of ethnic minorities had gradually come to be seen as the needs of 'ordinary' consumers of housing.

This recognition of ethnic minorities as normal community residents is a crucial development. However, the opening-up of subsidised housing to ethnic minorities was still rather selective in many respects. On the demand side, there was insufficient knowledge of how to deal with the housing market, and as a result the demands of ethnic minority people were on the modest side. Landlords, too, implicitly or explicitly assumed that old, low-cost housing would be the appropriate answer to ethnic minorities' housing demands. So on both the supply side and the demand side the stakes began low. These low stakes were also a result of economic developments. The gradual breakthrough of minorities into the housing market took place at a time when policies that directly targeted low-income households were apparently nearing their end. The products of urban renewal were becoming more costly, while the buying power of the households that needed them was falling. Ethnic minorities in the cities were especially hard hit by the economic malaise. Unemployment had traditionally been low amongst them, but it now soared to many times that of the ethnic Dutch urban population.

All the same, the time when minorities always lived in substandard housing was a thing of the past. In 1994, 40 per cent of the tenants of the 20,000+ renovated dwellings in Rotterdam belonged to ethnic minorities.

Case Study 1: Migrants in the decision-making process At the transition between the first and the second phases in post-war urban renewal, it was increasingly evident that the problems in the old city districts demanded an integrated plan of action. In the period we have described as Building for the Neighbourhood, special project organisations were set up for this purpose. They played a critical role in a network made up of representatives of municipal services, private institutions, and community organisations. Minority groups were not part of these networks in the beginning. In this case study we describe the entry of minorities into the networks, and the marginal position they acquired there. Our account is derived from research by Van der Pennen (1988) on the functioning of such networks in Rotterdam in the late 1980s. Before looking at the position of minorities we shall first examine the role played in the networks by traditional community organisations, which were composed of ethnic Dutch residents supported by community workers and architects.

As a result of a process of professionalisation and institutionalisation, the role of the community organisations gradually shifted from that of an initiating or opposing party to that of an 'ordinary' coalition partner. In the same period, the social composition of the group they were supposed to be representing altered substantially. The proportion of the population from Turkey and Morocco grew, and in some neighbourhoods and streets these 'minorities' even became a majority of the population. Because their participation was far below that of the rest of the neighbourhood population, the Rotterdam city authorities decided to hire special 'migrant community workers', who were sent on secondment to community organisations and municipal services. The objective of this experiment, which has since been terminated, was to increase ethnic minority participation in urban renewal, both by making it more accessible to them, and by ensuring that their interests were heard in specific urban development projects. Broadly speaking, we can distinguish here between individual and collective approaches to migrant community work. The former method consists mainly of individual support to ethnic minority residents during urban renewal projects, particularly through home visits. In the collective method, emphasis is on helping them to organise around issues surrounding their housing and living conditions.

Various studies have revealed that such migrant community workers were insufficiently integrated – or totally unintegrated – into the urban renewal policy network (see, e.g., Van der Pennen, 1988). One major explanation for this lies in the character of that network. By the time such workers appeared on the scene, it had become more or less a closed circuit, in which one dominating approach to problems, and a standard success formula, had evolved. As 'newcomers', the workers did not succeed in acquiring a fully fledged structural position in this institutional network. Recently appointed, and differently educated, these officials had undergone a briefer and 'deviant' socialisation process in urban renewal circles in comparison with their ethnic Dutch colleagues. The differences were so great that they never gained any clear place in the network, in contrast, for example, to the architects who advised neighbourhood residents. The backgrounds and frames of reference of the ethnic Dutch officials in the network were usually similar. They either knew or knew of each other because they had been to the same college or studied for the same degree. The migrant community workers generally were not part of the 'informal project organisation'.

It could, of course, be enlightening for new voices to be heard inside such a 'closed' system. But the migrant community workers were not successful enough in getting their insights translated into policy, or even getting them onto the agendas of the project organisations. Certainly every day they ran into problems which were worthy of attention from planning teams, but which in practice did not usually get that far. In many cases they sought solutions to problems of individual migrants by acting outside the organisations. The experiences that migrant community workers gained, by personally talking to and helping people, rarely penetrated to the higher policy levels, and therefore were not translated into concrete policy. Migrant community workers did not speak the same language as other officials in the network. They were unable to translate problems specifically affecting migrants into the dominant terminology of the policy network. Or conversely, their contributions were viewed through the lens of the dominant urban renewal ideology. Because these workers were also accustomed to different forms of interaction and different work relations, tensions and conflicts were not uncommon. An important factor here is that the migrant community workers tended to feel isolated. The fact that their jobs were temporary, as part of an experiment, may also have been relevant.

The communication problems between migrant community workers and their network colleagues are thus not only a question of 'misunderstanding' one another. In that case language and training courses could solve many of them. But it is also a question of different traditions of thinking and approaches to work which are not accepted in the existing network. An official who fails to conform to the dominant set of norms will remain an outsider. The positions in the network have become ossified. New voices like those of the migrant community workers go unheard.

The next question is whether the various actors/officials responded constructively to the contributions of the migrant community workers.

This has been investigated in the case of community organisations. These organisations consisted then, as now, of a cadre of professionals and ethnic Dutch residents, who despite their pretensions represent only part of the ethnically mixed population of the neighbourhood. In fact, ethnic minorities in all their diversity are not represented in this cadre at all (Van der Pennen, 1988). We can distinguish different patterns of reaction to the arrival of migrants.

For some community organisations it more or less went without saying that ethnic minorities would develop relatively autonomous initiatives. Like Heinze et al. (1987), we would characterise their approach as 'a strategy of offensive integration'. Typically, these community organisations see themselves in no way as the sole representatives of neighbourhoods. As institutions they concern themselves with the problems in the neighbourhood, and consider it important for residents of different ethnicity to do likewise. In such situations, mutual co-operation develops, and migrant community workers receive ample support in their efforts to set up ethnic minority interest groups for the neighbourhood. At least, the community organisations are open to initiatives by the workers to mobilise ethnic minority residents. Some variations can be seen in the degree to which minority activities are integrated into these organisations or remain relatively autonomous.

In most community organisations, by contrast, we encountered a so-called 'strategy of wait-and-see tolerance'. While the workers did get the room to put their ideas into practice, they received no real support at all in doing so. The problems of migrants were seen as an exclusive task of migrant community work. In situations of this type, considerable tensions were known to arise in the course of time, because both sides turned out to harbour expectations that the other side failed to fulfil.

Although none of the community organisations we studied displayed open hostility to the migrant community workers, a hostile attitude on the part of some participants was evident on occasion. Several workers reported that certain people had openly stated that minority residents, and the community workers representing their interests, were not welcome in their organisations.

This all shows how difficult it is to give minorities a voice in urban renewal activities. Even during this period, when citizens' participation was a central goal, the position of minority groups remained marginal in policy networks. Measures benefiting minorities were therefore more likely to be inspired by their Dutch defenders than by active participation by minorities themselves. It is difficult to predict how the situation will evolve further in this respect. There are conflicting developments. On the positive side, more and more people from ethnic minorities in the Netherlands have obtained a good education, which can facilitate involvement in policy networks. One outcome of the process of governmental decentralisation is that ethnic minority residents have gained more opportunities to participate in politics and government. But the negative side of the coin is that in recent years institutions involved in urban spatial planning and public housing have grown less keen on citizens' participation, especially when strategic decisions are at issue. Furthermore, there is a distinct tendency in the Netherlands to replace policy which targets specific groups with policy aimed at alleviating deprivation in general.

Urban revival Even though urban renewal achieved clearly visible results in the course of time – large sections of the old city districts were vigorously refurbished – the social problems proved intractable. Over the years the incomes of residents of urban renewal districts had lagged further and further behind those of people elsewhere in the city (SCP, 1992, pp. 212-213). The exodus of long-time (ethnic Dutch) city residents continued, and so did the influx of ethnic minority people. This process of segregation and concentration, it was presumed, impeded the integration of the latter, thereby further entrenching their social disadvantage. Building for the Neighbourhood had not succeeded in halting this longer-term process.

Another consequence of the vigorous policy interventions – which were concentrated one-sidedly on housing – was that small-scale, informal economic activity in urban renewal districts had severely slackened in these years (Brand, 1988). This had limited the opportunities that marginal groups in such neighbourhoods had to build up their own existence without recourse to social security.

In the third phase of urban development, starting around 1990, the city in its totality came back into the picture. Development was increasingly part of a discourse on economic revitalisation. It was now argued that urban renewal must serve to make the most of the city's economic potential: to make the city not only attractive, but also accessible and habitable for the more prosperous. 'Distinction', 'premium locations' and 'prestigious' are among the catchwords in the plans now being implemented. The purpose of urban renewal is to convince the business world, preferably international firms, that Dutch cities are good places to locate. To this end, the city has to create an ambience which, in both design and available facilities, is attuned to the wishes of future users. Economic revitalisation is deemed essential if Dutch cities are not to fall behind in the tough international competition between cities, regions, and markets.

In contrast to the mono-functionality of the 1960s and the one-sided focus on subsidised housing in the 1970s, metropolitan projects are now to include a mix of commercial, residential, and recreational functions. New challenges have been formulated, to ensure an adequate response from public and private organisations involved in making decisions about the design, or management, of dwellings, and their surrounding environment. Increasing efforts are made to interest private market parties – banks, investors, construction companies – in investing in the urban building and housing market, alongside non-profit private organisations such as housing corporations. The intended outcome is a public-private partnership with the ultimate goal of optimising the city's economic potential.

Not only has the position of trade and industry in urban decision-making altered since the 1970s: so has the position of ordinary citizens. The latter had played an active role in the previous urban renewal process, but with the revamping of urban renewal, citizens have been assigned the role of consumer. Policy is focused on the local and regional housing market, which is organised more in keeping with market principles. The general public are customers in this market, especially when it comes to the more expensive accommodation to be built in the future. These shifts, which occurred in the early 1990s, have not made it easier to secure adequate housing for ethnic minorities. Increased immigration has given rise to a good deal of 'new demand' for low-rent accommodation, on top of the persisting demand, from households of longer-term residents, for large and still affordable dwellings.

This has all served to make ethnic minorities more dependent on developments in the existing housing stock. Policies on rents, housing quality, and housing distribution, are crucial factors that determine whether enough good-quality dwellings become, and remain, available for this range of market demand. A large part of the new housing being constructed is in higher price categories, and much of the replacement housing, built after the demolition of poor-quality buildings, is subject to stricter requirements of price differentiation in renewal areas. As a result, many of the evicted tenants will move to pre-existing housing rather than to the new dwellings.

In the mid-1990s, the relationship between urban renewal and ethnic minorities is viewed primarily from the angle of whether undesirable social segregation occurs which derives from income and ethnicity. If this is found to be the case, corrective measures are sought through urban renewal methods. In metropolitan areas, minorities have become ordinary consumers with just a few special traits. These derive not only from their low incomes and high unemployment, but also from the fact that relations between them and the ethnic Dutch population of the cities are under tension. The economic malaise has not spared the latter either, and support for right-wing extremism has grown since 1990. Looking into the future, we can predict that over one third of the urban population in 2005 will have an ethnic minority background (first, second or third generation).

Social policy Alongside the present optimism about the revitalisation of the cities and a revaluation of urbanism, there is pessimism about the spread of poverty and decay. Both of these phenomena can be observed, and they form the two poles of urban development – a process of polarised growth. Economic revival always benefits certain sectors of the economy and the population, while others fall victim to stagnation and decline. The dividing lines between different groups in the cities seem to be sharpening. We have documented this empirically earlier in the chapter. We have pointed there to the mismatch between supply and demand in local labour markets. New, high-grade jobs are created as a result of urban revitalisation, but they draw workers from outside, while a large share of the present city residents see their chances steadily diminish (see also Van der Wouden, 1996). Fostering revitalisation of the city must not result in a simultaneous sharpening of the divisions within it. Economic revitalisation and social problems are two sides of a coin. But solving those problems is in itself not an aim of the present 'city renewal policies'.

In the early 1990s, 'social renewal' became the new catchword for local social policy. In the light of the developments arising from urban renewal, social renewal was especially intended to avert a split within cities. Keen attention is now paid to the everyday living environment in the cities (see Van der Wouden et al., 1994; Van der Pennen et al., 1996). The neighbourhood is the starting point in most cases. The local authorities have stimulated co-operative projects at that level to combat long-term unemployment, to lower school dropout rates, and to improve neighbourhood management. Municipal departments are there to provide expertise. According to local policy memoranda, these departments are to pick up on ideas from residents or other interested parties and to ensure they are carried out, irrespective of sectoral boundaries (see Van der Wouden et al., 1994).

The new social policies are to help combat effects which central government feels are threatening to the social climate in neighbourhoods and city districts. Dirty streets, petty crime, conflicts that smoulder or erupt between local groups, and concentrations of disadvantaged groups, are all adverse influences on the social climate in some districts, notably the chronically deprived areas. Policy to improve the situation of migrants in the Netherlands is pursued not only in certain sectors such as public housing, the labour market, and education. There is also an overall minority policy, instituted in the 1970s. Specifically targeted policy was needed because migrants were settling permanently in Dutch society. They were unevenly dispersed through the country, and concentrated in the large cities and towns, and this began to be a source of problems. Minorities appeared to have the characteristics of an underclass, with on average a low socio-economic position.

As soon as migrants had settled in the Netherlands, interest groups and a large number of organisations sprang up in their communities, with the aims of cultivating group identity and strengthening mutual solidarity. National government policy initiated at that time was primarily in support of such initiatives.

In the early eighties, however, it became clear that such assistance was failing to improve the poor socio-economic situation of the migrants (Minorities Memorandum, Lower House of Parliament 1982/83). New policies were then formulated, aimed at an equitable distribution of goods in society and at making the various societal institutions more accessible to migrants. Members of minority groups (migrants) were to participate in employment, education, and housing, on an equal basis. The policy amounted to a virtual war on poverty. Specific measures for migrants were to be taken when considered necessary. A later policy evaluation, however, established that these compensatory policies for migrants had borne little fruit. In the early 1990s, policy to further the integration of migrants received a new impulse, or took a new turn. More than was previously the case, it was now based on mutual obligations – migrants were more than just an object of care – and on the need for migrants to exert themselves more towards improving their lot.

Looking back at the developments just described, we can conclude that policy to counter urban decline and social exclusion runs along two tracks. Urban development measures are taken with an eye to economic revitalisation, and social policy measures are taken to enhance the position of migrants. This is illustrated in the case study below.

Case Study 2: The case of the Bijlmermeer: a strategy to combat decline and exclusion The Bijlmermeer district of Amsterdam is one of the Netherlands' oldest and best-known areas of ethnic minority residence. It has an undeniably bad reputation, and is universally regarded as a problem area. It is of interest as a case study because it illustrates how migrants are dependent on the least attractive parts of the housing market, and also because a host of measures are currently being undertaken to address the problems there. From the angle of urban revival, drastic redevelopment is being promoted, such as the building of new dwellings for higher-income groups. In the sphere of social policy, a multitude of projects have been initiated to tackle the social problems facing the district. Before going into the problems and the interventions now being made, we will first outline the history of the area. It is inextricably bound up with two issues: the original development plan of the district, and the large-scale immigration of Surinamese people in the first half of the 1970s. Our description is based largely on the following publications: Mentzel (1989), Brakenhoff et al. (1991), and information material from the Vernieuwing Bijlmermeer project bureau (1993-96).

The creation of the Bijlmermeer and its present predicament are above all the story of an urban development fiasco. From the early 1960s, the Amsterdam local authority was drawing up plans for a large-scale housing development south-east of the city. Working-class families from the older city districts were to find homes there, which, unlike their previous ones, were suited to the requirements of the modern day and age (see Brakenhoff et al., 1991; Gemeente Amsterdam, 1962). The local authority envisaged an estate with plenty of greenery, massive high-rise housing blocks, and a stringent separation of functions: an estate built according to CIAM principles. That part of the district finished between 1968 and 1975 amply satisfied these standards. Especially striking are the 30 massive galleried buildings about ten storeys high. Because of their extreme length they were given a honeycomb shape to break the monotony. Each was surrounded by generous amounts of greenery. After 1975 low- and medium-rise housing was also built, but the Bijlmermeer remained essentially a high-rise estate with, in 1991, 15,000 flats in the high-rise buildings, 7,000 in smaller buildings up to five storeys, and 2,600 low-rise dwellings. Some 80 per cent of the total housing stock consists of subsidised rental dwellings (see Brakenhoff et al., 1991).

It was already clear by the late 1960s that the flats in the new estate were going to be too expensive for the original target group (Brakenhoff et al., 1991). Without any market research, and without any adaptations to the plans, a new target group was advanced. This was the same group of middle-class families being targeted by the suburban residential areas that were arising in large numbers from the 1970s onwards at a slightly greater distance from the city. The middle class naturally showed little interest in the Bijlmermeer, and moved en masse to suburbia. From the very beginning of the Bijlmermeer, accommodation was found there by many households who had a weak position in the tightly regulated housing market in the rest of the city (Bolte, 1982). Construction faults came to light in some buildings at an early stage, and by 1974 two problem buildings were already plagued by substantial vacancy rates.

On the eve of the independence of the former Dutch colony of Surinam in 1975, many of its inhabitants emigrated to the Netherlands, and many of these settled in Amsterdam. Their position in the housing market was weak, and they were therefore forced to accept the least popular flats, which at the time were in the Bijlmermeer. Despite early dispersal policies intended to redirect the influx of immigrants away from particular neighbourhoods, a significant proportion of the Surinamese population ended up in the Bijlmermeer (see Brakenhoff et al., 1991).

In the years to follow, other groups of immigrants would join them (see Table 3.2). They were mostly from the Dutch Antilles or from African countries like Ghana, or were asylum-seekers. Since many of the Surinamese are so-called Creoles (descendants of African slaves), the Bijlmermeer now has a concentration of people with an Afro-Caribbean cultural background.

It should be noted in relation to Table 3.2 that great differences in ethnicity exist between the high-rise and low-rise sections of the district; there are also differences within the high-rise complexes. Some of the latter are still largely the domain of older ethnic Dutch people, while others became purely Surinamese in the course of time.

The Bijlmermeer contended with management problems from the very start, and these have multiplied over the years. They include burglary, drug dealing, vandalism, high tenant turnover, over-occupation, and rent arrears (see e.g. Brakenhoff et al., 1991). Besides crime and the plethora of management problems, the high unemployment rate demands attention.

Various attempts have been made to do something about the problems of management, crime, and social deprivation. Buildings were refurbished, and the district was targeted with measures to alleviate deprivation. Such interventions proved insufficient to correct the adverse course of development in any fundamental way. In 1993, a sweeping package of measures was introduced to address the spatial structure and management of the Bijlmermeer. It was partially funded by the European Commission's URBAN programme. The most crucial measure is a drastic redevelopment scheme that includes the demolition of several of the high-rise buildings (containing about 3,000 dwellings), the renovation of 8,000 other flats, and the sale of some 750 flats to tenants. Fourteen parking garages are to be demolished, because of their negative effect on people's sense of safety.

The demolition of non-condemned residential buildings on such a scale is an operation hitherto unheard-of in the Netherlands. The land it yields is to be used primarily to build single-family dwellings. The local authority and the housing association hope this will create a residential environment that will be valued more highly than the present one. They hope both to attract new, more affluent residents as well as to retain the current residents who have paid work, in an effort to break the present mechanism by which people leave the district as soon as they find a job. Building for groups who are socially stronger, instead of for the sitting tenants, would not have been possible several years ago, and it illustrates that Building for the Neighbourhood, the guiding principle from the 1970s, has been abandoned once and for all. It has made way for the principle of Building for the City.

Many of the present interventions are hence not aimed at the needs of the current Bijlmermeer residents. Nevertheless, the construction of single-family dwellings will clearly also benefit some of the Surinamese residents, enabling them to improve their living conditions and still be near family and friends. At the time of writing, the demolition phase of the redevelopment plan has been partially completed and the construction of single-family houses has begun. While it is not yet possible to evaluate the impact of the programme, a few comments can be made. Great interest has been shown in the homes being built, so the plan is successful in this respect. Since much of the interest comes from outside the district, it is not yet known how many dwellings will ultimately be occupied by Surinamese people from the neighbourhood. A further question, of course, is what will happen to the remaining blocks of flats. Will they be gathering points for the deprived, or will they develop more favourably? And what consequences will the redevelopment have, in the longer term, for other Amsterdam districts? After all it is quite conceivable that some of the nuisance, especially that connected with narcotics dealing, will move to other areas.

In addition to the structural redevelopment, other measures have been adopted in the sphere of social management and the easing of deprivation. One of the premises of the redevelopment project is that spatial and social renewal must go together, and that people currently living in the area must also profit from the renewal. Thus, in the context of social management, more surveillance has been introduced, by hiring extra caretakers and watchkeepers, many of them recruited from among the numerous unemployed tenants. There are also experiments with video monitoring. Surveys among tenants have shown that such measures have a positive effect on their perceptions of safety (Boumeester, 1996).

We have already pointed to the high rate of unemployment in the Bijlmermeer district. This has proved an intractable problem. There are projects which provide training and create subsidised jobs. In the context of national policies to alleviate deprivation, over 300 heavily subsidised jobs have been created in the Bijlmermeer for surveillance, environmental maintenance, and socio-cultural work. The construction company contracted for the redevelopment programme also committed itself to hire unemployed people. But it is debatable whether this approach can do much about the problem. The high unemployment rate is tied to the gap between supply and demand on the labour market: many unemployed residents have little education, and there is a great shortage of unskilled jobs. This is why unemployment levels remain so high, and why Bijlmermeer residents have benefited so little from the strong growth in employment opportunities in the immediate area. They fail to find work in the offices and industries that have located around them, and the employees of such firms do not yet seem inclined to move to the Bijlmermeer.

In concluding this case study, we should point to an unanticipated effect of the present revitalisation process. A conflict in the Bijlmermeer district council, about the allocation of URBAN funding, has resulted in an ethnification of political decision-making. A so-called Black Caucus, made up of black representatives from different parties in the council, has recently become active. Unique for the Netherlands, this situation has provoked a good deal of internal animosity within the political parties in the council. The future will tell how political decision-making will evolve further. It is unlikely, however, that ethnic minorities in other districts will follow the Bijlmermeer example. The differences between the diverse minority groups in those districts are too great.

Conclusion

The position of ethnic minorities in the Netherlands is not very encouraging. For some twenty to thirty years now, the Dutch have been accustomed to living with Turks, Moroccans, Surinamese, Antilleans, and other ethnic groups from around the world. Despite this, the 'inclusion' of those groups in the labour market, the housing market, and the education system, has not yet been achieved. An 'ethnic question' still exists today, though the situation is vastly different from that in a recent past.

Step by step, ethnic minorities have taken root in Dutch society. In the housing market, they have worked their way up from lodging houses, through substandard flats, to better-quality, more modern, rental accommodation, and dwellings refurbished in urban renewal projects. For many landlords they are now ordinary customers, but this improvement in their housing conditions comes at a time when government policy depends more on the actions of market forces. Groups in a weak position cannot rely on as much policy protection as previously. The more prominent role of the income factor in the housing market forms a particular disadvantage for ethnic minority people who have been so hard hit by unemployment since the 1980s. Thus, the pressing issue of minorities is being felt at a time when social goods, such as housing, continue to be scarce, or are available only at higher prices. In other words, while minorities are now getting a larger piece of the pie, the pie itself has become smaller and dearer. The role of minorities in decision-making has suffered a similar fate. Just as minorities start to assume a more 'normal' role, the authorities are attaching less importance to citizens' participation.

The two-pronged strategy of economic revitalisation and social renewal presents both opportunities and threats to ethnic minorities. Opportunities are arising for small-scale enterprise, and for new forms of involvement in neighbourhood management. But policy on labour and housing could also develop in an exceedingly market-oriented direction. In that case, patterns of exclusion will continue to manifest themselves for groups with low levels of schooling and with modest bargaining positions in the housing market.

Table 3.1 Minorities,[a] by size of municipality, 1992

(percentages calculated horizontally)

	four big cities	towns over 100,000 inhabitants	towns under 100,000 inhabitants
Turkey	37	21	41
Morocco	48	14	38
other Mediterranean countries	34	16	50
Surinam	57	13	31
Dutch Antilles	31	21	47
non-industrial countries (nic)[b]	37	17	46
total population	13	14	73

a Based on country of birth of at least one parent.
b Immigrants from Asia, Africa or Latin America, when not part of other categories and when officially registered.

Source: CBS, excerpt from Register Count at request of SCP.

Table 3.2 Bijlmermeer population according to ethnicity

					other foreigners			
Surinamese	Antillean	Turkish	Moroccan	South European	non-industrialized countries	industrialized countries	Ethnic Dutch	Total
18,090	3,940	1,024	1,084	956	12,384	2,830	13,402	53,710
33.7%	7.3%	1.9%	2.0%	1.8%	23.1%	5.3%	25.0%	100.0%

Source: O+S, 1995.

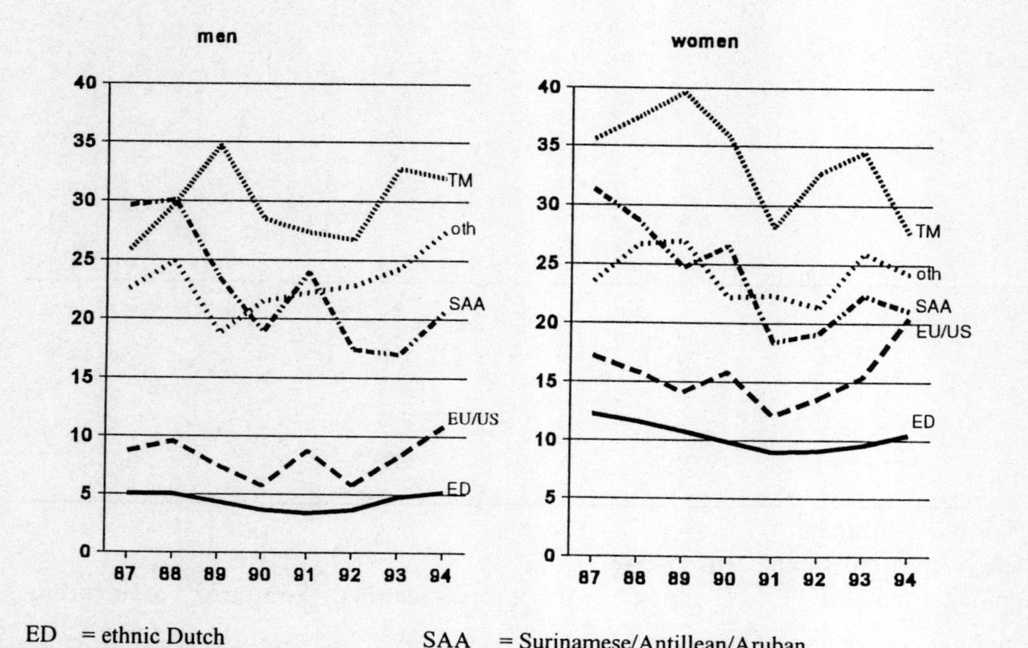

ED = ethnic Dutch
TM = Turkish/Moroccan
SAA = Surinamese/Antillean/Aruban
EU/US = European Union/United States
oth = other origins.

Figure 3.1 Unemployment by ethnic origin, 1987-1994 (percentages)

Source: CBS (EBB '87-'94).

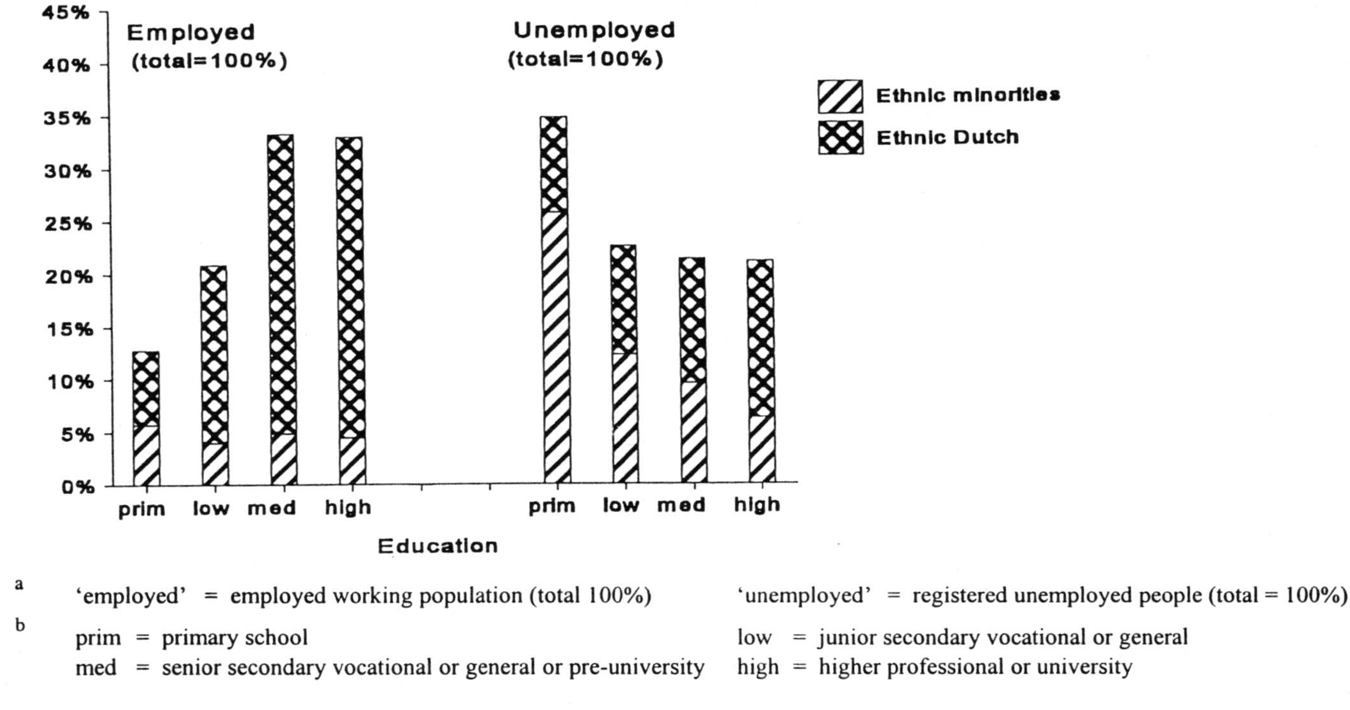

a 'employed' = employed working population (total 100%) 'unemployed' = registered unemployed people (total = 100%)

b prim = primary school; low = junior secondary vocational or general; med = senior secondary vocational or general or pre-university; high = higher professional or university

Figure 3.2 Employed and unemployed[a] residents of the four big cities by educational level[b] and ethnicity, 1992

Source: CBS (EBB '92), adapted by SCP.

References

Bolte, W. and Meijer, J. (1982), *Van Berlage tot Bijlmer*, Sun, Nijmegen.

Boumeester, H. and Wassenberg, F. (1996), *Video voor veiligheid? Effecten van camerabewaking in de Bijlmermeer*, OTB, Delft.

Brakenhoff, A., Dignum, K., Wagenaar, M. and Westzaan, M. (1991), *Hogebouw, lage status. Overheidsinvloed en bevolkingsdynamiek in de Bijlmermeer*, Universiteit van Amsterdam, vakgroep sociale geografie, Amsterdam.

Brand, A. (1988), 'Probleemcumulatiegebiedenbeleid en deelname van langdurig werklozen aan stedelijke vernieuwing in Rotterdam', in *Integrale stadsvernieuwing: nieuwe perspectieven.* Congresbundel, R.I.W., Delft, pp. 25-39.

Gemeente Amsterdam (1962), *Om de toekomst van 100.000 Amsterdammers*, Amsterdam.

Heinze, R.G., and Olk, T. (1987), *Non-profit and self-help organizations in the welfare state: competition or cooperation?* Paper at International symposium on the non-profit sector and the modern welfare state.

Kassarda, J.D. (1985), 'Urban change and minority opportunities', in Peterson, P.E. (ed.), *The new urban reality*, The Brookings Institute, Washington DC.

Kassarda, J.D., Friedrichs J., and Ehlers, K.E. (1992), 'Urban industrial restructuring and minority problems in the US and Germany', in Cross, M. (ed.) *Ethnic minorities and industrial change in Europe and North America*, Cambridge University Press, Cambridge.

Kassarda, J.D. (1995), 'Urban industrial transition and the underclass', in Wilson, W.J. (ed.) *The ghetto underclass*, Sage.

Kloek, J. (1992), *De positie van allochtonen op de arbeidsmarkt*, Centraal Bureau voor de Statistiek, Heerlen.

Kloosterman, R.C. (1994), 'Amsterdamned. The rise of unemployment in Amsterdam in the 1990s', *Urban Studies*, vol. 31, pp. 1325-1344.

Lewis, O. (1966), *La Vida*, Random House, New York.

Martens, E.P., van Roijen, J.H.M., and Veenman, J. (1994), 'Minderheden in Nederland', *Statistisch vademecum 1993/1994*, SDU, Den Haag.

Mentzel, M. (1989), *Bijlmermeer als grensverleggend ideaal*, Delftse Universitaire Pers, Delft.

Niesink, W., and Veenman, J. (1990), 'Achterstand en achterstelling op de arbeidsmarkt', in Veenman J. (ed.) *Ver van huis. Achterstand en achterstelling bij allochtonen*, EUR-instituut, Rotterdam.

O+S (Amsterdam office for research and statistics) (1995), *Jaarboek 1994, Amsterdam in cijfers (Amsterdam in numbers)*, Amsterdam.

Pennen, A.W. van der (1988), *Migrantenwerk in de stadsvernieuwing. Een studie naar de betekenis van een participatiebevorderende maatregel voor allochtonen*, ROV/Universtiteit Leiden, Leiden.

Sociaal en Cultureel Planbureau (SCP) (1992), *Sociaal en Cultureel Rapport 1992*, Rijswijk/Den Haag.

Sociaal en Cultureel Planbureau (SCP) (1995), *Rapportage Minderheden 1995. Concentratie en segregatie*, Rijswijk/Den Haag.

Smit, V. (1991), *De verdeling van woningen: een kwestie van onderhandelen*, Technische Universiteit (dissertatie), Eindhoven.

Tesser, P.T.M., van Dugteren, F.A. and Merens, A. (1996), *Rapportage Minderheden 1996*, Sociaal en Cultureel Planbureau, Rijswijk/Den Haag.
Veenman, J. (1991), *Allochtonen op de arbeidsmarkt. Van onderzoek naar beleid*, Stichting voor Strategisch Marktonderzoek, Den Haag.
Wilson, W.J. (1987), *The truly disadvantaged: the inner city, the underclass and public policy*, University of Chicago Press, Chicago.
Wouden, R. van der (1996), *De beklemde stad. Grootstedelijke problemen in demografisch en sociaal-economisch perspectief*, Rijswijk/Den Haag.

4 Ethnic minorities, urban renewal and social exclusion in Italy

PAOLA SOMMA

Ethnic minorities and local government

The 'special statute' regions

In a cross-national context, Italy provides a paradigmatic case of long-standing localism and centralised power. The contradiction is only an apparent one. Italy achieved political and administrative unification later than other European nations, and compared with other European countries, a sense of shared national identity has never been strong. Italian national politics has been aggregative rather than integrative, based on the pursuit of particular interest more than the commitment to a common set of values and the adoption of a collective world view (Agnew, 1995).

Acknowledging the existence of unsealed cleavages in the formation of the national state, and of minorities capable not only of claiming a separate ethno-linguistic identity but also of asserting their right to independence, the Italian Constitution of 1948 guaranteed a special self-government statute to five of the country's regions. Initially there were four: Val d'Aosta (French speaking), Trentino-Alto Adige (German speaking), and the islands of Sicily and Sardinia, where separatist tendencies needed to be neutralised. Creation of the fifth autonomous region, Friuli-Venezia Giulia, was delayed until 1963 owing to difficulties concerning Trieste. Initially, regional autonomy was only put into practice in these 'special statute' regions, with specific linguistic or cultural characteristics. Extension of regional autonomy to the 15 'ordinary' regions had to wait until 1970.

In Val d'Aosta and Trentino-Alto Adige the populations identify with the dominant ethnic groups of neighbouring states, and differ from other Italian regions in language, historical memory, and culture. They have generated genuine political parties, which govern the region and send representatives to the State Parliament in Rome.

Ever since their institution, the special statute regions have enjoyed substantial privileges; and relationships with the 'Italians', especially in Alto Adige, have often given rise to conflict. These two minorities have not only achieved recognition of their identity but also receive favourable treatment - they have been called ethnic elites - in exchange for forgoing their aspirations to secede from Italy. By contrast, other smaller and weaker ethnic groups, such as the Slovenians in the north-east or the Arberesche in the southern province of Cosenza, endure rather harder conditions.

Bolzano, a divided city

A city where conflict between ethnic communities is reflected at a territorial level is Bolzano, where ethnic claims have significantly impinged on distributional questions at the municipal level, and where town planning schemes have consistently set out to separate the Italian language population from the German speakers. The first provisions following the annexation of the city to Italy, in 1919, employed mass migration as one of the means to reinforcement of the Italian presence, and its concentration, in the newly expanding areas. The wish to Italianise the city was openly proclaimed during the Fascist period; and under several plans, attempts were made to expand the suburbs, in order to reduce the weight of the German component of the population, which was mainly concentrated in the centre of the city.

In 1910 the Italian-speaking inhabitants numbered 3,600, and there were 26,600 German speakers. By 1939 the ratio was reversed, with 36,600 Italian speakers out of a total population of 58,000. The Italian presence, due to the thoroughgoing colonising operation undertaken during the two decades of Fascist domination, was further accentuated in 1939 because of the exodus of German language inhabitants that followed the pact between the Italian and German governments providing for 'options'. The move-out made clearance of the ancient heart of the city easier, and the administration simultaneously urbanised new development areas in an attempt to reach 100,000 inhabitants. In 1943 there were 50,700 Italians and 13,800 Germans.

As well as demolishing many central districts inhabited by industrial workers and other low-income families, and zoning residential areas on a class basis, the Fascist regime also prepared planning schemes designed to implement ethnic segregation. In the immediate post-war period the administration carried on with its policy of infinite urban growth, on the expectation of continuous immigration, the objective now being to reach a population of 150,000.

But after the Second World War the prevailing nature of the relationship between the two communities changed, and consolidation of the supremacy of the German group became the main criterion. The autonomous Province of Bolzano adopted an anti-industrial, anti-urban type of development model, based on the weakening of the central role of the municipal administration and of its socio-economic vitality. As well as reducing the population of Bolzano in favour of the other municipalities of the province - in this connection it should be remembered that while 75 per cent of the population of Bolzano municipality were Italian, in the province as a whole 75 per cent were German - the policy placed a high priority on agriculture and conservation in the mountain communities. This kind of development guaranteed favourable conditions for the conservation of its traditional forms of socio-economic organisation, but this positive result is to a large extent the consequence of a politico-social policy to ensure the dominance of the German ethnic group.

The restrictive discipline that governed settlement policies, especially in the residential and industrial sectors, achieved the unquestionably positive result of restraining urban expansion; but the parsimonious use of land resources penalised the 'more urban and weaker' social strata (who were mainly though not exclusively Italian), leaving their housing needs unresolved.

With the coming into force of the new autonomous status of the region in the 1970s, conflict between a minority and the State was transformed into conflict between the different linguistic groups. The climax of this process of change came with the introduction in 1976 of a quota system, which marked the beginning of a policy of retaliation on the part of the German group in relation to employment and housing problems. The so-called 'ethnic proportional' criterion provides for distribution of civil service jobs, welfare benefits, public finance, and subsidised housing, in proportion to the size of the groups officially recorded during the census.

The 1961 census recorded the language used by the householder and in 1971 the language group of every person covered by the census. The purpose of these data, however, was mainly statistical. The 1981 census, though, contained for the first time a mechanism for measuring ethnic groups, based on a declaration of ethnic attachment required of every person covered by the census, the purpose being to identify each person's ethnic status and consequent rights. This mechanism, which institutionalised the conflict, gave rise to protests - 2,600 people refused to declare their ethnic affinity - but it is in force, and Bolzano remains a divided city.

Town planning, at both theoretical and practical levels, plays a by no means secondary role in creating divided cities, though it is not always easy to find land use proposals and projects in which separation features as a declared objective. Even when not explicitly declared, the intention seems to be to create ethnically homogeneous and tendentially segregated zones. This can be perceived if population movements within the city are compared with the town planning schemes that followed them.

The Province of Bolzano is the only part of Italy where the census records each individual's ethnic attachment. It offers confirmation of the general phenomenon according to which minority groups with greater power and resources will themselves seek separateness. The clearest example is that of South Africa, where the purpose of apartheid was to exclude the majority of the population. It is necessary to clarify our terms therefore: not only the concept of ethnicity, but also that of minority.

With the increase in social polarisation, and the probable intensification of migratory flows from the poorer countries, belonging to a minority will less and less be the consequence of a mere quantitative ratio between different population groups.

The history of migration

Italy: from emigration to immigration

During its existence, Italy is estimated to have 'exported' over 26 million people (King, 1987). Migration was particularly intense in the first decades of the century, and in the post-war period it continued to be a massive phenomenon. Internal migration has been as significant as emigration abroad. In ten years, from 1951 to 1961, more than 15 million people moved from one part of Italy to another (Montani, 1966). In the last twenty years a new phenomenon of immigration flows has developed quickly, moving from the southern coast of the Mediterranean Sea towards the southern European countries. Italy has become increasingly attractive for immigrants. In 1973, for the first time, surveys of Italian migration flows showed that whereas 101,771 Italians left their homeland, 125,168 had decided to return.

Some attribute the halt in the net outward flow of Italian citizens to internal factors: basically the country's emergence from a phase of economic backwardness; others place more importance on the fact that many European countries had adopted a policy of replacing Italian labour with less qualified workers from further afield; and still others draw attention to Italian workers' relative lack of competitiveness in Europe. All agree, however, that the beginning of the 1970s marks the end of a cycle. 1974 was a turning point in Europe, with the official cessation of immigration. A transformation in the number of immigrants, together with profound changes in the structure of these groups and in the nature of their relations with the society welcoming them, took place. Even internal immigration came to a final halt, despite the continuing gap between unemployment rates in the various regions.

The same period marks the first significant examples of immigration of non-Europeans, who were initially concentrated in economically and geographically marginal sectors, and who later became increasingly present in more advanced regions and activities. Already in the previous decades, newspapers had reported the presence of nuclei of African and Asiatic domestic workers in the larger cities, and the growing number of Yugoslavians crossing the border daily to work in Italy, and of Tunisians employed on a seasonal basis in the fishing industry on the coasts of western Sicily.

For as long as the entrance of immigrants into the labour market was limited to filling gaps in sectors using manual and unskilled labour, the debate remained at a local level. The question attracted national interest only when the presence of immigrants came to be considered as a source of competition threatening the employment opportunities of the national labour force. The news, in 1977, that several North African workers had been taken on by foundries and china factories in Reggio Emilia and Modena, prompted a series of concerned articles and surveys in the national press, and may to some extent be considered as a turning-point in ways of looking at the problem.

The government ordered Censis to carry out a nationwide survey.[1] Censis chose a certain number of observation zones in which to concentrate its field surveys, from which it would extrapolate estimates for the remaining areas as to the degree to which their economic structure might attract foreign workers. Four economically specific areas were chosen for the survey: Milan, as representative of a great metropolitan area; Veneto/Friuli/Venezia Giulia, because of its characteristic productive fabric of small and medium-sized business concerns; Emilia Romagna for its role in the Summer tourist industry; and Sicily, for its fishing industry and dock labour (Censis, 1978).

New arrivals of non-European immigrants, most of them clandestine, continued throughout the 1980s. From the end of the 1970s, clandestine immigration increased in almost all European countries, but in none so much as Italy, where the majority of the immigrants now resident arrived clandestinely, and were subsequently able to get their papers in order through successive amnesties. This element, together with the fact that at least until 1989 official surveys failed to record the presence of most immigrants, should be remembered when data concerning the presence of immigrants in Italy is being considered (see Table 4.1).

The reduction recorded for 1989 in the table is really only apparent, because it is due to the records being updated. The increase between 1986 and 1990 was concomitant with the possibility of regularisation given to immigrants by two national laws (n. 943/1986 and n. 39/1990). The most recent data, updated to August 1995 and published by the Ministry of the Interior, speak of 809,936 non-EEC immigrants who are properly registered and in possession of a residence permit. Other statistics and estimates from various sources speak of approximately 2 million immigrants, 1 million of them legal and the other million clandestine.

The present situation

The incidence of non-European immigrants is still low, on the whole, in Italy. The phenomenon remains of relatively modest dimensions when compared with other countries, but a large number of different nationalities and ethnic groups are represented among the immigrant population in Italy (Table 4.2). The five largest groups account for over 30 per cent of the immigrants. Not only is the phenomenon of foreign immigration into Italy very heterogeneous in nationality and type of migrant, but it has impacted on the various Italian regions in different ways (Montanari and Cortese, 1993a) (Table 4.3). Over 50 per cent of immigrants reside in 10 provinces.

Several different types of immigration can be identified in Italy, distinguishing on the basis of individual characteristics (sex, age, education), motivation (economic, political, cultural), length of stay intended (short, medium or long), occupation, and whether or not immigrants have legally valid permission to stay.

Arabs from North Africa are the most numerous immigrants in Italy. They tend not to settle, they move around a lot, and there is a high turnover rate. Apart from certain 'historic' settlements, like the Tunisians of Mazara del Vallo, they tend to concentrate in the large cities, where they are able to find work in the lower-paid reaches of the service sector. Basically the group comprises young, unmarried males, but growing numbers of family nuclei are being reported.

Immigrants from the Philippines, from the Creole culture Cape Verde Islands off the west coast of Africa, and from certain states of western India, tend to be women, who find domestic work in the medium and large cities. Immigration flows from Sri Lanka, Eritrea, South America, and El Salvador, tend to include a substantial number of political refugees. The last few years have seen the rapid increase of immigrants from south of the Sahara; there were previously very few. This movement is the most disorganised and fragmented. From the early 1990s the stream of migration from Yugoslavia and Albania has been increasingly large. Finally there are the Chinese, the only immigration movement that has given rise in certain cities (Milan in particular) to thoroughgoing encapsulated communities, a term which anthropologists use to define a homogeneous closed group with a clearly defined territory and a strong cultural identity.

Today, immigrants find work in the secondary band, or on the edges of the labour market. The vast majority work in the informal service sector (home help, hotel and restaurant staff, porterage in markets and transport, street trading, etc.) while the rest are employed in agriculture and industry. According to 1994 Ministry of Works statistics, just over 21,500 non-EEC immigrants have properly registered jobs in agriculture, while it is difficult to gain an accurate picture of the situation as regards clandestine immigrants, who frequently move from one region to another as work opportunities present themselves.

The rapid and substantial increase in immigrant flow has been accompanied by a qualitative change. In particular, there has been a disproportionate increase in the number of socially problematic immigrants; the labour market is having greater difficulty absorbing them, and with the consequently increased unemployment, they take up irregular and unstable occupations, find it practically impossible to obtain housing, and drift increasingly into various forms of social marginalisation and even common crime (Bergnach and Sussi, 1993).

Immigration policy

An increasing number of sources on issues of migration, ethnic relations, and racism, have become available in Italy in recent years (Fondazione Agnelli, 1990; Delle Donne et al., 1993). Most studies concern the immigrant presence and its characteristics in the various regions of the country (Dell'Atti, 1990; Natale, 1990; Reyneri, 1991; Minardi and Cifiello, 1991). Another area of research deals with institutional and social change, from the drafting of new legislation, to the emergence of examples of intolerance. Public policy has exhibited tensions which reflect social conflict. For example, on the one hand the Law n. 39 provided for reception procedures that were intended to guarantee equal social rights to all citizens, whether Italian or foreign; on the other, it laid down strict rules governing entrance into the country.

During the 1990s, public opinion with regard to immigrants has gradually become less sympathetic, and institutional attitudes have hardened, from the point of view both of legislation and of the practical implementation of immigration policy. The identification between immigrant and worker has been weakened by the economic and class-identity crises. The strengthening of the identification between immigrants and the poor promotes forms of assistance and dependency, and a distinct identity, which in turn encourages the emerging of social conflicts, and a dangerous spreading of xenophobic attitudes.

After an initial provision in 1986, which enabled a certain number of immigrants to legalise their position, the most important law in this connection was introduced in 1989. Since then the immigration problem has been discussed several times in Parliament. On each occasion, however, what occurred was a head-on clash between two ideologies, rather than a debate about concrete proposals and solutions. Also, the recent proposal that immigrants should be given voting rights in local elections is being countered by the argument that 'the privileges of non-EU immigrants must not be paid with money due to Italian pensioners'. At present, immigrants have no political power, and it is most unlikely that they will gain power by organising themselves on an ethnic basis.

The speed of the process that transformed Italy into a country with a significant immigrant presence quickly revealed the country's general unpreparedness to deal adequately with the situation. Immigration is still seen as an evolving and transitory phenomenon. Instead of developing effective reception strategies, the government has concentrated on trying to limit new arrivals. During 1994 the authorities issued expulsion orders for 64,000 foreign citizens, and 12,000 expulsions were actually carried out. Taking this approach a stage further, in 1995 the government decided to use the armed forces to prevent the landing of Albanian immigrants in Puglia, a phenomenon controlled by the Turkish, Chinese and local Mafia.

Territorial distribution and the 'reception centres' for immigrants

The spatial dimension becomes more important in the second stage of the migratory cycle, when immigrants are joined by other members of their families. In the case of certain European countries, where post-war economic expansion set in motion considerable population shifts and immigration flows, there are studies which prove the existence in certain cities, especially in Germany, France, and Holland, of ethnically defined zones.

In general, studies of ethnic minority settlement in major cities depend upon evidence derived from population censuses. Changes through time cannot be undertaken as process studies, but only as comparisons between given dates. Usually, geographical areas, and not individuals, are the research units. Moreover, these were descriptions of cases considered in isolation, and it is only in recent years that scholars have undertaken comparative studies designed to develop an interpretative model which can take account of the phenomenon of segregation in European cities.

In Italy, much has been written on population movements from one part of Italy to another, on territorial imbalance, and on the living conditions of southern Italians who have moved to the great cities of the north (Fofi, 1963; D'Alasia and Montaldi, 1975).[2] By contrast, there have so far been very few studies of the territorial implications of the new migrations that have affected not only the big cities but also relatively rural parts of the country (Tosi, 1993).

After an interesting study of Tunisians in Sicily (Guerrasi, 1978), the first organic attempt to describe the settlement mechanism of a substantial number of foreign workers was carried out in Milan at the beginning of the 1980s (Caputo, 1983). There is, however, no more general consideration of the relationships between non-EEC immigration and the reorganisation of urban space (Somma, 1991).

Attention is still focused on the immigrants' living and occupational conditions (Birindelli, 1991), but there is no reliable, up-to-date information about their territorial distribution (Altieri and Carchedi 1992; Canup, 1992; De Filippo and Morlicchio, 1992), nor about possible example of segregation. The position is not assisted by the fact that the Italian population census does not collect information on ethnic groups.

In Italy, local administrations have considerable responsibility for territorial policy, and thus play an important role in this field: they draw up the municipal town planning scheme, and designate the areas for subsidised housing developments and public services. In 1990 they were also given, by the Law n. 39, the task of, and funding for, preparing reception centres for immigrants. The municipal authorities used the funding in five main ways:

- residential service centres;
- purpose-built reception centres, usually using prefabricated structures;
- adaptation for residential use of disused, publicly-owned buildings;
- procurement, and possible restructuring, of dwellings through payment of rent, or specific arrangements between owners and immigrant tenants;
- purchase of dwellings on the market, followed by possible restructuring.

The decisions which some local authorities have taken, concerning the designation of certain areas and of certain buildings for settlement by immigrants, seem to have taken no account at all of the consequences of such decisions on the general organisation of the city, and the administrators have confined themselves to comments confirming the 'spontaneous' nature of settlements in certain run-down areas. Study of the siting, construction, management (and now dismantling) of these centres enables several situations to be outlined.

The most interesting cases are:

- the large metropolitan areas of Turin, Milan and Genoa, where every phase of the process aroused considerable controversy and in some cases clashes between different interest groups.

  Milan City Council was the first to try to deal with the housing situation. In 1982 it decided to allow immigrants to rent publicly-owned dwellings. Eritreans have been the principal beneficiaries of this policy, having been allocated about 500 dwellings. But Milan City Council is now controlled by the Northern League, and the climate is very hostile to immigrants.

  In many cases the lack of facilities to accommodate migrants has forced foreign workers to resort to temporary, precarious solutions. The former Pantanella factory in Rome was occupied in 1987, and the Cascina Rossa in Milan offered shelter to thousands of people who lived in subhuman conditions (Montanari and Cortese, 1993b).

- a number of medium-sized cities in Emilia Romagna and Tuscany, such as Modena, Parma, and Pistoia, which have traditionally been governed by left-wing administrations, and where there has been an attempt to integrate immigrants into the society as a whole.

  In the province of Parma, for instance, according to statistics from the local police office, there are over seven thousand legally registered immigrants. About half of them come from the Magreb countries on the Mediterranean; the remainder include other Africans and refugees from Eastern Europe. They usually find seasonal work in agriculture and in the construction sector. The municipality of Parma has set up a first-help centre, but the main support is still given by the church and the volunteer organisations (Trufarelli, 1989).

- a number of small and medium-sized towns in the Veneto: Treviso and Vicenza for example, at the moment governed by the Northern League, but where there is a strong presence of religious associations, often working in co-operation with trade union organisations.

A reconstruction of the attitudes and approaches of these cities to the settlement problems of non-EU immigrant groups should yield additional insights towards an evaluation of the wider scale territorial policy decisions.

The advantage of adopting, as an indicator, the spatial opportunities afforded by a weak group, is that this provides not only a point of comparison but also a warning, if it is true that 'nowadays, many see Europe itself becoming inward-looking and racially exclusionary with the enforcement of EC (European Castle) immigration policies' (Greed, 1994, p.150).

Urban policy

At no time since the Second World War has Italy had an explicit urban planning policy. To say that Italy has lacked a real urban development policy does not mean that the cities have not constituted an important element in Italian policy-making, but the origins of such policies lie in successive governments' financial and fiscal strategies, and in measures to do with public works and regional and local autonomy, rather than in urban planning as such.

Features of this implicit policy include the inadequate, or in any case unobeyed, regulations intended to control land use, and the habit of regarding the unauthorised occupation of, and building on, public land as quite normal (and the *a posteriori* legalisation of such abuse). In this programmed uncertainty, and taking into account the fact that literature presenting empirical material on specific urban planning episodes is somewhat sparse - only in recent years has work been published on specific cities (Indovina, 1993) - certain considerations can nevertheless be made.

At the national level, there is no doubt that the most significant and consistent feature of a whole succession of laws and provisions concerned in some way with action with regard to our cities is the fact that they were always 'emergency' measures.

At first the emergency was justified by the need to rebuild the war-damaged urban centres, and to provide some form of housing for the great mass of people moving from the south of Italy to the industrial cities of the north. Then the justification became the need to respond to so-called natural disasters (which in fact were often provoked or aggravated by the lack of adequate regional policies). There are numerous examples of this phenomenon, many of them tragic: from landslides caused by building work in dangerous areas, to floods consequent upon the lack of any kind of hydrogeological control, and to the now emblematic case of the reconstruction of areas devastated by the earthquake in the region of Naples some years ago.

The policy of proceeding by declaration of an emergency, with the disbursement of massive public funds and no organic planning instruments, has had seriously perverse effects on expenditure and waste of public resources, the loss of democratic control over decision-making, and environmental damage (Campos Venuti and Oliva, 1993). At the same time however, the procedure accrued enormous advantages to the building contractors, financial groups, technical experts, and politicians involved, and it achieved widespread *de facto* acceptance.

Though the collections of legislative provisions promulgated as emergency measures seem not to have been based on an unified design containing any identifiable consciously adopted planning policy, taken together they do represent a 'mode of governing' that became further reinforced during the 1980s. When deregulation became rooted to the point where even the most modest reforming experiment was out of the question, the emergency situation became the indispensable prerequisite for any kind of urban planning action. Now, however, it was no longer a case of responding to unforeseen calamities but rather of programming large-scale spectacular events.

At this stage, the recourse to an emergency, often one artificially created with the help of specific publicity campaigns, served to accelerate the entrenchment of so-called 'bargained planning'. This involves abandoning any attempt at constructing a system of established rules and regulations which are the same for everyone. Instead, decisions regarding urban transformation projects are negotiated directly between those parties having decision-making and finance-allocating powers.

'Bargained planning' has had even more serious consequences than those of traditional building speculation and traditional illegality. Initially, violations of town planning schemes were, at least theoretically, illegal; now it is suggested that a town plan should be flexible and adaptable to the changing requirements of the private sector. The contradiction, in certain senses objective and positive, between public control of urban transformation, and private construction of urban growth, has been resolved by eliminating the control (in the sense of the forceful exertion of an independent public design), and allowing the private sector increasingly to assume *de facto* the rights and functions of government.

The second element characterising action as regards the city (and which may to some extent be considered the inevitable complement to the 'emergency' approach) was the adoption of a series of Special Laws for specific cities. The institution of the Special Law is by no means a recent development - and indeed the first town planning law to be passed in Italy was promulgated in 1885 as the 'Law for Naples' - but since the Second World War the expedient has been used with considerable frequency. There have been Special Laws for many cities, from Reggio Calabria to Venice, and from Matera to Ancona. The financial balance has been disastrous in all cases, and the achievements far below expectations. In part this may be considered a consequence of a system of redistribution of resources which has so far been based on the disbursement of funds by central government to the cities concerned, and therefore on the existence of pressure groups seeking the allocation of additional resources; and in part it is disturbing evidence of a lack of attention to the problems of the country as a whole, and of the prevalence of evaluation criteria which have nothing whatsoever to do with urban planning.

While the relative weight of the various participants was changing, and private interests were more and more openly taking the place of the public interest in territorial government, local authorities were also modifying their attitude. Perhaps the most important transformation concerned cities in which left-wing local governments had been brought to power following the great movement of widespread social and trade union struggle during the 1970s. There was great hope at the time, that after the plans of the 1950s, designed to consolidate the process of land accumulation begun during the post-war reconstruction period, and in effect making it possible to build on all land within the municipal boundaries, innovative elements might now be introduced into the structures of the cities and their management.

In the majority of cases the response was inadequate, and modernisation became a synonym of 'large-scale works' or 'large-scale investments'. The presentation of projects unconnected with any plans, which became common during the 1980s, not only corresponded at a local level to the emergency and special case strategies pursued by central government, but also represented a fundamental feature of the privatisation of town planning.

An appreciation of the way the great private financial groups assumed the planning role, which many municipal administrations had proved unfit or unable to perform, is essential to an understanding and evaluation of the planning policies of recent years, not only in the great cities such as Rome, Milan, Florence, Naples, Genoa, and Venice, but also in the country as a whole. The changes brought about in relations between institutions and private, external, operators, by the establishment and growth of an operational practice that places greater store by spending capacity than by planning activities, can now be seen to have been crucial to the nature of the town planning policy choices exercised. Whether the actions proposed are more or less consistent with the plan is not particularly important: what matters is that certain operations should take off before others, and that their profitability should be guaranteed by the public sector.

In Naples, the left-wing Mayor, having proved his efficiency by clearing the city of beggars and immigrants on the occasion of the G7 Meeting, announced his intention of going ahead with several of the projects for the city centre, and with the redevelopment of certain industrial areas that had long been awaiting implementation. Private entrepreneurs had invested a great deal of energy and expertise in making up for the inertia of the municipal administration in these areas, and since 1986 have presented several grandiose redevelopment projects. The proposals do not confine themselves to the actual planning project but also include implementation procedures, defining the responsibilities of public authority and private operator, quantifying the finance required, and defining the amounts to be supplied from public and private sources. In effect this is a reversal of the roles normally performed by public and private bodies: the responsibility for defining planning decisions and setting the rules governing the operation falls to the private partner, while the job of the public counterpart is to legitimise the operation and disburse the funding required.

In most Italian cities the great development schemes are not preceded or followed up by any assessment of their environmental or social impact. The urban transformation process is guided by neither an overall design nor an adequate response to social needs, but rather by opportunities for the use of public resources in redeveloping available spaces, and by the emergence of particularly strong interests.

Urban renewal

Post-war clearance and reconstruction

The development of the capitalist system in Italy is closely connected to urban ground rents. The importance of the land ownership and building sectors in the development process is connected with other fundamental aspects of the country's economic system, including the simultaneous presence of highly developed and backward areas, technically advanced and obsolescent industrial sectors, and oligopolistic and almost competitive markets. These features are ascribed to the late-comer nature of the Italian economic system, and the land sector is therefore seen both as an index and as a consequence of 'delayed and partial' development.

Post-war reconstruction of central areas completed the clearance started by the fascist regime, and the destruction caused by bombing. During this period, several substantial speculative operations were able to take advantage of wartime destruction in the central areas of leading cities, using criteria that differed little from those followed in the previous Fascist period. In a certain sense, the reconstruction shows the extent to which the old mechanisms were also able to operate in the new situation of representative democracy, with municipal councils, mayors, parties, and trade unions.

In Rome, for example, the demolition of the buildings in Via della Conciliazione took place according to plans drawn up in the 1940s, almost as if to signal a cultural and political continuity with the previous regime. This had been responsible for building the so-called 'borgate', to house the throngs of Romans evicted from the centre of the city to make way for 'imperial' clearance and reorganisation schemes. The granting of tax exemption facilities to property owners and builders, and the use of extremely high building density indices, triggered a process of land speculation which drew fresh stimulus, once the reconstruction phase was over, from the growing concentration of immigrants in the metropolitan areas.

In the 1950s and 1960s, the massive amount of new building, and the relatively small incidence of renovation of the existing stock, were a consequence of the processes of population concentration in urban areas. In those years, public action focused on areas away from the centre, where urban landowners, property owners, and building contractors, were guaranteed unrestricted profits. The new, working class housing districts thus opened the way to private speculation.

In the south, too, where urbanisation took place without industrialisation, most new building generated a sort of oil-slick expansion of the outskirts of urban areas. In 1953, Italy had 37 million rooms for 47.5 million inhabitants, and by 1971 there were 63 million rooms for 54 million inhabitants. But at the same time, and particularly during the 1960s, the supply of housing available to lower-income families increasingly lagged behind demand, and this quickly became translated into insecurity of tenure, increased numbers of slum-dwellers and evictees, and the gradual split of the larger cities into two distinct zones: the suburban ghetto-housing estates, surrounded by shanty-town type accretions, and the gradually depopulating city centre. The expulsion of residents from city centres was assisted by the fact that the law regulated rent levels but not contract expiry dates.

Increasingly, thanks to an exemplary case of integration of private capital, urban land rent, and public capital, city centres became the only suitable places for prestigious economic and social activities, and the term periferia (suburbs) came to connote a range of diverse situations with an underlying homogeneity (Ferrarotti, 1970).

The historical centres

During the 1970s, Italy began to concern itself with urban renovation, and there was a change in attitudes to the existing housing stock and the old city centres, which often featured the widespread presence of monumental buildings interspersed among housing for a comprehensive range of social classes. When the outskirts of towns became saturated, and older districts regained their appeal, central areas, which had grown steadily in value during the unrestricted exploitation of the suburban zones, became the subject of renewed attention. This gave rise therefore to a dual speculation: on the one hand in agricultural areas immediately outside the towns, which became reserve zones for future urbanisation; and on the other in the existing housing stock, hitherto occupied by a substantial part of the less well-off section of the population, and by part of the blue- and white-collar working class. The historical sections of cities showed a spread process of gentrification, and faced massive land-use changes. They became the concentration point for the tertiary activities.

Urban renewal encountered considerable resistance, however, and often took different forms from those found in other European countries. In many cities, base committees were formed to try to stop the expulsion of low-income families. They often joined forces with workers' action groups and trade unions, and, especially after the 1975 local elections, when left-wing parties were able to get into power in most local governments, gained the support of many administrations. Though they sometimes managed to slow down the phenomenon, the organised movements in favour of housing rights, and against 'deportation', were ultimately unable to stop the expulsion of inhabitants from areas earmarked for redevelopment. In the Garibaldi district of Milan, for example, the inhabitants mobilised, and in 1968 presented an alternative spatial reorganisation project to the one proposed by the municipal administration, which had encouraged private real estate speculation. Nevertheless the entire population of about 40,000 was moved out, and their place taken by new inhabitants with the means to live in a luxury district (Crosta, 1972).

At the end of the 1970s, the halt to large-scale migratory movements, connected with the reduced demand for labour on the part of industry, caused a territorial redistribution of the population within the more urbanised areas, and highly selective migration patterns. In this period, population movement mainly involved moving from one part of the metropolitan area to another, and gave rise to a process of polarisation, with heavy losses from the central social classes. This trend has intensified in recent years, at the same time as the ongoing processes of economic reorganisation in the country, thus demonstrating that although segregation in consumption and in everyday life, and controlled integration in the interests of production, are the main characteristics of the settlement patterns of the subordinate classes in any historical phase, spatial separation and discrimination may take more or less accentuated forms.

In some cases, 'natural' disasters have added an element of urgency to renovation work designed to ensure the preservation of a gradually deteriorating part of the historical-environmental heritage, thus hastening the process of population replacement and the conversion of buildings to other uses. In Naples, until the 1960s, the older districts of the city centre featured a socio-professional structure comprising (sometimes in the same building) a component of unstable proletariat typical of an urban subsistence economy; the craftsman with his workshop attached to his home; the upper-middle classes; and, more rarely, the working classes. In the 1970s, there was, first, a steady exodus of those who enjoyed a stable income. Then, in 1974, the outbreak of cholera served as an opportunity to speed up the process. Though the epidemic struck most seriously in the suburban public residential zones, attempts were made to link the need for hygiene improvements to proposals for redevelopment of the central zones. The municipal authority adopted a town planning scheme that provided for extensive demolition. This scheme aroused opposition from many citizens, and the Ministry of Public Works intervened to reduce the amount of destruction. The event did, though, set in train a process of population movement, that later involved further social strata following the 1980 earthquake. Indeed, for the implementation of the Special Public Housing Plan of 1981, which provided for the construction of 20,000 new dwellings, entire families were shifted to the edges of the urban and metropolitan areas, and to segregated conditions quite different from the socio-professional mix that had characterised the old districts of the centre. The 'cleared' centre thus became the focus of grand projects for new office buildings, and the Mayor, at the head of a left-wing governing council, became a fervent supporter of the need for Naples to 'follow the example of Baltimore'.

Urban segregation

The dual role of the local authority - to satisfy housing needs (with maximum political advantage and minimum economic outlay, searching for low-cost land), and sustaining the private building sector by commissioning public housing estates to fill that land - accentuated the differences and the separation between the central and peripheral parts of the city (Ferrarotti, 1970; Cerasi, 1973).

After the initial post-war phase, which featured similar criteria to those used previously (with small and medium-sized projects), the later self-contained developments of the INA Casa programme were sited in isolated areas on the outer edges of the suburbs; here they performed a pioneering urbanisation function for the adjacent areas, ready for them to be taken over by speculative housing projects.[3] These developments, clearly aimed at the poorer social strata, in fact helped to intensify their segregation.

In this phase the city centres functioned as reception areas for immigrants. These first settled in rundown areas, occupying housing that had been abandoned by the native population. Some suburban housing developments feature the presence of the complete range of medium-to-low social strata, but the immigrants live separately in dwellings built in the context of various Plans. The result is that, although the various classes and sub-classes live side-by-side in the same district, each social layer is confined to a particular type of building. Some blocks, for example, are accessible only to clerical workers and blue-collar workers, who are up-to-date with their welfare contributions; while the typical inhabitants of other buildings are the less well-off: tenants evicted from private rented housing, and former slum-dwellers.

Where there has been no, or only partial, renovation, central areas are now occupied by non-EU immigrants. The areas of the city centre where immigrants now tend to live are those occupied by immigrants from southern Italy during the post-1945 reconstruction.

Turin: San Salvario district[4]

In Turin, for a long period, urban planning and administrative management paid little attention to the presence of large-scale industry, thereby disregarding the effect of its development policy on the urban structure. Reconstruction and development took place without control, with land acquisition by estate agents and public housing developments being sited in marginal zones, this in turn resulting in marked spatial segregation of the population (Marra, 1985).

After an initial phase in which the immigrant population occupied the more dilapidated blocks and buildings of the city centre, it quickly overflowed into peripheral areas where building land was available and cheap. Furthermore the existence of natural barriers such as the River Po and the Dora, and the structure of the city, which is organised around a grid of straight-line traffic arteries and the north-south railway axis, all helped to make the various districts somewhat impermeable to each other.

The boom years saw the intensification of the industrial monoculture, and with the traumatic transformation of the immigrant population into a working class, Turin became increasingly polarised by comparison with the metropolitan area. When a new wave of immigrants entered the city in the early 1970s, the accumulated tension of the previous period exploded in the streets, factories, schools, and institutions. The demand for housing, which culminated in 1975 in the occupation of several buildings in the centre, helped to block a plan that would have meant the expulsion of most of the weaker productive and residential activities.

Speculative building and renewal started up again during the 1980s. Borgo S. Paolo, where Fiat tried in vain to build an office block during the 1960s, is an emblematic case of the social transformation of an old nucleus. The previous residential buildings were almost completely replaced by new medium- to high-cost developments built to service a vast commercial area, and this set off an uncontrollable process of change in the occupancy of the housing.

Of the 23,980 non-EU citizens with permits to stay in the province of Turin, 15,000 (1994) were registered as resident within the municipality of Turin. If neighbouring municipalities are also taken into account, the metropolitan area of Turin contains 20,000 non-EU immigrants with regular stay permits, representing approximately 2 per cent of the resident population. It is estimated that there are a further 6,000 to 8,000 who are not in possession of stay permits; but even then, with 2.8 per cent of the population accounted for by non-EU immigrants, the figures are much lower than in other major European cities, and far below the so-called 'critical threshold' (a very ambiguous but widely used concept).

San Salvario is a district in the centre of Turin, near the main railway station, with around 13,000 inhabitants. Being close to the station, it has traditionally been a centre of immigration, the initial point of arrival for workers attracted by the city's factories. San Salvario lies between a rich area, near the park of Castello del Valentino, and the poor district around via Saluzzo, with a high incidence of immigrants from southern Italy. With more and more foreigners settling here since the early 1980s, it has now become in effect a multi-ethnic district. The official incidence of non-EU immigrants rose from 3.2 per cent in 1991 to 6.2 per cent in 1995. But according to the estimates of the city administration, the actual number of foreigners in the area now exceeds 2,000, though many of them are not properly registered. At the beginning of the 1990s, with the growing numbers of foreigners additionally swelled as a result of racist reactions and displacements from other parts of the city, and with the arrival en masse of Nigerian prostitutes who altered the 'normal' market, there was an increase in social tensions, and periods of calm alternated with explosions of protest.

Periodically San Salvario occupies the front pages of national newspapers under sensationalist headlines - 'Blacks only in the street of the Bronx of San Salvario after 5 pm. The whites stay out of sight' - that evoke a false and in any case exaggerated image of breakdown (Romagnoli, 1995). The authorities treated the social conflict as a question of law and order. Following a torchlight demonstration by the tenants of a tenement block owned by one of the main speculators, where dozens of people lived crowded in insanitary conditions, the authorities cleared many flats occupied by non-EU immigrants, set up inspection posts, and increased the number and visibility of the police in the area, which has also been visited recently by the Minister of the Interior.

In actual fact, San Salvario does have some weak points, connected with the settlement of a 'marginal' section of the population, but it is no more a problem area than other parts of the city. It occupies an important position in the general context of the city, lying as it does between the main commercial areas, the railway station, and Turin's principal park. The quality of the buildings is also quite high, but they will deteriorate quickly if not subjected to maintenance work soon. There is therefore a real risk that the decay will become self-perpetuating, though it is difficult to quantify such a phenomenon.

As has happened in many other European cities, processes of degeneration, prompted by factors which are extrinsic to local society, tend to be followed by a mobilisation of public opinion, which then focuses on a contingent situation in which the immigrant population functions as a catalyst. In the case of San Salvario, it became clear that houses and shops were losing value, and in the words of the parish priest, 'if considerable improvements are not made in the enforcement of law and order, the inhabitants will begin to take matters into their own hands; people will put up with a lot, but when they see their property endangered they will rebel' (San Salvario, 1994).

In the opinion of some commentators, the attention given to the San Salvario case can be explained by the fact that it coincided with discussions in Parliament about the new Decree Law on immigration. The municipal authority commissioned CICSENE to conduct a study on San Salvario.[5] Their report draws attention to the fact that, while there certainly are several obvious problems (poor hygiene, squalor, drug-peddling), there are also a number of positive socio-economic qualities (e.g. a well-rooted and fairly diversified commercial presence, and healthy support for local associations) (CICSENE, 1996). The typical deterioration of historic buildings has not yet been followed by the usual process of restoration and gentrification. The building stock seems to be undervalued in real estate terms. It is not possible to demonstrate that speculative real estate interests are active in the area. Though the decline in property values is patently obvious, no single individual or interest has been identified as manoeuvring to take advantage of the situation. The appearance of the district is still homogeneous and coherent from an environmental and architectural point of view. It was built over a relatively short space of time. The negative connotations of the area are heavily dependent on subjective perceptions; the mere presence of groups of foreigners is taken as an indicator of physical and social decay.

The report, mindful of the fact that policies for intervening in complex areas must themselves be complex, suggests that priority should be given to upgrading the physical appearance and substance of the area, with projects that help to reverse the currently prevailing pessimism and desire to escape. The report suggests, too, the opportunity of fabricating a sort of 'quartier latin'. The effort to incorporate multi-ethnic diversity into central city redevelopment projects is a new phenomenon for Italy, but in other situations is a part of central city revitalisation. This trend is explicit in recent urban history, mainly in the United States, where traditionally ethnic places were perceived as undesirable districts and now, 'in a post modern developmental environment they have acquired a new sentimental salience' (Lin, 1995).

At the time of writing, the future of San Salvario is still uncertain; but marketing ethnic places and districts as commodities, and transforming immigrants into a spectacle of urban tourism and retail consumption, is probably an option taken into consideration by the local authorities.

Genoa: Borgo Prè district

Genoa's historic centre is the largest in Europe, and one which has deteriorated more than most. Several projects have been presented for its regeneration, but local residents were very reluctant, because an improvement usually means an influx of middle-income or upper middle-income residents moving into a neighbourhood. The main changes in its social composition coincided with the great economic transformations of the last hundred years. Thus, at the beginning of the century, the district was abandoned by the trading middle classes; after the Second World it received immigrants from southern Italy, and now, in the current phase of de-industrialisation, it accommodates the influx of immigrants from developing countries. In actual fact, the term 'accommodate' scarcely evokes appropriate associations; it would perhaps be better to say that an unspecified number of people have installed themselves in an area already full of poverty and hardship, and that exploitation of the vulnerabilities of the new marginal population has inevitably led to further privation.

Approximate estimates, made on the basis of information from various sources, suggest a changeable situation, with a fluctuating population of non-EU immigrants. Non-EU citizens in possession of stay permits in Genoa number about 12,000, but is it estimated that there are about the same number of clandestines. The great majority of them live in the historic and badly decayed Prè district, where the local population, ageing and economically very weak, is now around 30,000. The largest ethnic group is represented by 7,000 from Morocco.

Perhaps more than in other Italian cities, the building stock and economic fabric have gradually lost value, and maintenance of public spaces has increasingly been neglected. Tension has grown, exacerbated by irresponsible press campaigns based more on sensationalist headlines - e.g. 'The heart of Genoa, a black time-bomb' - than on an analysis of real problems, which are mainly due to economic reorganisation that has forced a severe cutback in port and industrial activities, and not to the presence of immigrants, which might accelerate but which cannot be the root cause of the decay.

In 1989, a group of inhabitants prepared a questionnaire, and assembled the replies they obtained into a White Book that revealed the extreme hardship of life in a district characterised by marked squalor and social alienation, and where traditional commercial and craft activities seem to have gone into irreversible decline (Genova: libro bianco sul centro storico, 1990). Though it confused cause and effect - the White Book took no account of the fact that the immigrants had settled in a district that was already in a state of decline - the initiative had the effect of starting a process of housing rehabilitation. This in turn led to the setting up of an association called ARCA (Association for the rebirth of the city centre) which compiled a second report, a Green Book (Genova: libro verde sul centro storico, 1992) which recorded the willingness of the inhabitants to become actively involved in their district. In the third and final phase, the association drafted proposals for a planned upgrading of the district, and submitted these to the public administration.

In the meantime, the episodes of intolerance and violence continued, and police operations increased, including round-ups of immigrants. Finally, in 1993, violent street riots occurred, and after the police's intervention the number of immigrants has been reduced. After the partial 'cleansing' of the area, the association obtained public funding for an 'integrated rehabilitation programme' for a block which was not very big but was in a pivotal position. The owners undertook to restore the buildings, the municipality to repair the pavements and urban furniture, and the public utility companies to renovate the sewers and the water and gas supply networks.

The project received publicity in several important journals, and was presented as an exemplary experiment in 'renewal through participation'. In the meantime, no action was taken to solve the problem of housing for immigrants. Now, at last, renewal is taking place.

Final considerations and directions for further research[6]

Many European cities with a large immigrant population are characterised by the appearance of ethnic districts in which the ethnic group in question is so densely settled that it develops its own relatively autonomous social identity. In general it is the disrepair and rundown state of the housing which lies at the root of voluntary concentrations of disadvantaged ethnic minorities. Sooner or later however, these settlements end up being considered as a social blemish and a possible source of conflicts, or they are taken up by the land or housing market and eliminated. Action in such enclaves may take different forms: from demolition and rebuilding, to restoration and consequent raising of rents; but in all cases the result is the moving away of the inhabitants.

The transformations which have affected the main cities of Europe over the last few years offer extensive evidence of various town planning mechanisms that have resulted in segregated communities, though there is a prevailing tendency to consider disadvantaged social groups in general rather than ethnic minorities, who are on the whole the most frequent victims of discrimination. One hypothesis is that the presence of immigrants is tolerated by the authorities in order to accelerate the process of degradation.

Given the limited dimensions of the immigration phenomenon in Italy, urban renewal has obviously not had the same ethnic implications here as it frequently has in other countries. Urban renewal in Italy has mainly been associated with socio-economic rehabilitation, gentrification, and the expulsion of weaker categories: low-income workers, pensioners, etc. However, there must be a serious reappraisal of what regeneration encompasses, and, especially, who it is intended to benefit.

Ethnic segregation and ethnic discrimination affect people who are not Italian citizens. There is a generally accepted tendency to allow immigrants with valuable human resources to come to Europe, but to find a means to stop the unaccepted mass immigration of lower-qualified people, refugees, and asylum seekers (Straubhaar and Zimmermann, 1993). In reality, however, the phenomenon of migration does not so much raise new problems as aggravate existing ones. It is no coincidence that in the southern parts of the country immigration is seen in terms of the existing problems associated with work and employment, while in the north it is seen as a problem of reception, housing, and the provision of services.

The increasing deregulation will worsen the position of the workers. Social services will be reduced. It is doubtful that familial and voluntary institutions, even in the richest areas of the country, will have sufficient capacity and resources to prevent serious forms of social distress. In certain cases, the municipal administration has responded to these emerging issues by resorting to the traditional welfare facilities set up for people with no fixed abode, and by co-operating with the voluntary services (Allasino et al., 1995); but on the whole, intolerance and racism are increasing (IRES, 1992). The growing presence of a multi-ethnic population has implications for the use, production and supply of urban space and services. Consequently, useful insights can be acquired by studying the activity of local authorities, especially at a municipal level.

Social disparity is not peculiar to any particular historical period but occurs in all forms of social organisation. Nevertheless, industrial society provided a more fertile breeding ground for the struggle for equality than did previous forms of social organisation. The distinctive feature of post-industrial society, on the contrary, is that it actually plans for inequality. Social inequity is no longer an accident that can and must be eliminated, or at least alleviated; it is now an objective that is specially pursued. The extolling of competition and localism involves the introduction of mechanisms that undermine the traditional forms of communal aggregation, and which, by encouraging the pursuit of individual priorities, accentuate social disintegration. The result is that new forms of division appear beside the old ones, and the city increasingly becomes composed of many minority enclaves that are ever more difficult to combine into a balanced mosaic.

If European cities are not yet 'crazy quilts' like the great metropolitan areas of America, it is because some forms of redistribution still survive; but the general trend is towards the accentuation of differences. It is not easy to measure inequality, but residential separation is a significant indicator, and the spatial dimension remains a crucial element in establishing the degree to which a city is divided. Throughout the history of modern town planning there has been a constant debate about whether different social groups should be mixed or kept separate, to what extent, and according to which spatial patterns; but now the concept of the poorer classes as dangerous, traditionally used to justify the great clearance and urban renewal projects of the last century, has once again become an explicit criterion of territorial organisation.

Notes

1 Censis (Study Centre for Social Investment) is a foundation, created in 1964, which performs research in the socio-economic field. Most of the studies are commissioned by institutions, particularly ministries and public bodies.

2 Useful reading for an understanding of urban changes in Italy after 1945 includes the following books:
Cederna, A. (1956), *I vandali in casa*, Laterza, Bari;
Romano, M. (1980), *L'urbanistica in Italia nel periodo dello sviluppo,* Marsilio, Padova;
De Seta, C. and Salzano, E. (eds) (1994), *L'Italia a sacco*, Editori Riuniti, and *Urbanistica,* Roma.

3 The INA Casa (National Housing Institute) programme was launched in 1949. Strictly speaking, it was primarily a plan to provide assistance for the unemployed rather than a plan to build new housing. It was financed through compulsory contributions made by workers and companies, and the dwellings that were built were allocated by lot. The programme was abolished in 1963 - but contributions are still being deducted from employees' wage packets!

4 Indirect information sources are used to construct case studies. No ad hoc research has been carried out, and the studies are based on material already available, mainly newspaper articles and documentation dossiers.

5 CICSENE is a non-profit-making body that operates in the field of international co-operation. It is based in Turin, and is particularly sensitive to issues relating to habitat, including the immigration phenomenon and the new forms of 'social demand'.

6 It is not easy to identify a research programme in a situation which is changing all the time, and which is invariably treated in emergency terms. One day it is a question of the Italian Navy sinking a ship laden with Albanians; the next the traffic in Nigerian prostitutes; and the next again the arrival of the Kurds, and so on. Each problem

occupies the front pages of the newspapers for a few days and then is relegated in favour of some other dramatic event.

At the same time, the growing hostility to refugees that one senses in the air makes it all the more necessary to pursue initiatives, even partial ones, that may improve our knowledge and understanding of the phenomenon.

Perhaps the subject of urban renewal is less suitable than it is in the other countries under consideration as a way of focusing on the mechanisms of social exclusion based on ethnic prejudice; a more useful approach may be through the more general question of the degree to which settlement opportunities are accessible to immigrants.

It might be interesting to start to deal with the question by identifying positive situations: situations where families and groups of immigrants have managed to become established (not necessarily integrated) and where their presence is not perceived as a danger; where schools function with a high percentage of non-Italian children, and hospitals with 'coloured' doctors and nurses; in short, where the arrival of foreigners has not given rise to pitched battles but rather to new forms of economic and social organisation.

Table 4.1 Numbers of properly registered non-EU immigrants from 1970 to 1995

1979	124,109
1980	198,871
1981	218,596
1982	241,671
1983	247,464
1984	260,551
1985	274,587
1986	293,349
1987	397,274
1988	457,124
1989	361,465
1990	632,527
1991	717,571
1992	778,254
1993	834,451
1994	831,129
1995	809,936

Source: Ministry of the Interior.

Table 4.2 Different nationalities present in Italy

	1990	1994	1995
non-EEC immigrants	781,138	922,706	
Morocco	80,495	92,617	94,237
former Yugoslavia	30,121	89,444	51,973
Tunisia	42,223	41,105	40,454
Philippines	35,373	40,714	43,421
Albania	2,034	31,926	34,706
	----------	----------	----------
	190,246	295,806	264,791
Senegal	25,268	24,615	23,953
Egypt	20,211	21,230	21,874
Romania	7,844	20,220	24,513
Brazil	14,555	19,589	22,053
China	19,237	19,485	21,507
Poland	17,201	18,929	22,022
Sri Lanka	13,214	18,689	20,275
Somalia	9,660	16,325	17,389
India	11,412	13,336	14,629
former USSR	6,701	13,329	15,426
Ghana	11,655	12,646	12,550
Argentina	13,116	10,681	10,494
Ethiopia	12,249	10,145	9,895
Peru	5,385	8,721	10,025
Colombia	5,615	8,253	9,626
Iran	15,022	8,038	7,956
Dominican Rep.	5,000	7,848	8,652

Source: Caritas.

Table 4.3 Properly registered immigrants by province, in 1994 and in 1995

	1994	1995
Rome	143,151	158,181
Milan	123,769	134,461
Naples	29,461	30,776
Turin	23,980	24,241
Vicenza	21,800	24,321
Florence	20,621	23,208
Palermo	18,562	18,032
Verona	13,952	13,765
Catania	13,656	13,376
Brescia	13,781	15,472
	----------	---------
	422,723	455,833

Source: Caritas.

References

Agnew, J. (1995), 'The Rhetoric of Regionalism: the Northern League in Italian Politics', *Transactions of the Institute of British Geographers*, vol. 20, no. 2, pp. 156-172.

Allasino, E., Baptiste, F. and Bulsei, G.L. (1995), 'Gli incerti confini: politiche per gli immigrati e politiche di lotta all'esclusione sociale a Torino e Lione', *Polis*, vol. 9, no. 1, pp. 23-44.

Altieri, G. and Carchedi, F. (1992), 'L'immigrazione straniera nel Lazio', *Inchiesta*, vol. 22, no. 95, pp. 29-39.

Bergnach, L. and Sussi, E. (ed.) (1993), *Minoranze etniche ed immigrazione*, Angeli, Milano.

Birindelli, A.M. (1991), 'Gli stranieri in Italia: alcuni problemi di integrazione sociale', *Polis*, vol. 5, no. 2, pp. 301-312.

Campos Venuti, G. and Oliva F. (eds) (1993), *Cinquant'anni di urbanistica in Italia. 1942-1992*, Laterza, Bari.

Canup, A. (1992), 'L'immigrazione straniera in Lombardia', *Inchiesta*, vol. 22, no. 95, pp. 15-28.

Caputo, P. (ed.) (1983), *Il ghetto diffuso*, Angeli, Milano.

Caritas (1997), *Dossier Statistico*, Roma.

Cederna, A. (1956), *I vandali in casa*, Laterza, Bari.

Censis (1978), *La presenza di lavoratori stranieri in Italia,* Censis, Roma.

Cerasi, M. (1973), *Città e periferia. Condizioni e tipi della residenza delle classi subalterne nella città moderna,* CLUP, Milano.

CICSENE (1996), *Problematiche di un 'quartier latin'. San Salvario-Torino*, CICSENE, Torino.

Crosta, P. (1972), 'I processi di rinnovo urbano e le implicazioni per l'operatore pubblico e privato nella situazione italiana', *Archivio di Studi Urbani e Regionali*, vol. 3, no. 1, pp. 79-117.

D'Alasia, F., and Montaldi, D. (1975), *Milano. Corea. Inchiesta sugli immigrati*, Feltrinelli, Milano.

De Filippo, E., and Morlicchio, E. (1992), 'L'immigrazione straniera in Campania', *Inchiesta*, vol. 22, no. 95, pp. 40-49.

Dell'Atti, A. (ed.) (1990), *La presenza straniera in Italia. Il caso della Puglia,* Angeli, Milano.

Delle Donne, M., Melotti, U. and Patilli, S. (1993), *Immigrazione in Europa. Solidarietà e conflitto*, Dipartimento di Sociologia dell'Università di Roma, Roma.

De Seta, C. and Salzano, E. (eds) (1994), *L'Italia a sacco*, Editori Riuniti, and *Urbanistica,* Roma.

Ferrarotti, F. (1970), *Roma da capitale a periferia*, Laterza, Bari.

Fofi, G. (1963), *L'immigrazione meridionale a Torino*, Einaudi, Torino.

Fondazione Giovanni Agnelli (1990), *Italia, Europa e nuove immigrazioni*, Fondazione Agnelli, Torino.

Genova: libro bianco sul centro storico, (1990), Sagep, Genova.

Genova: libro verde sul centro storico, (1992), Sagep, Genova.

Greed, C. (1994), *Women & Planning*, Routledge, London.

Guerrasi, V. (1978), *La condizione marginale*, Sellerio, Palermo.

Indovina, F. (ed.) (1993), *La città occasionale. Firenze, Napoli, Torino, Venezia*, Angeli, Milano.
IRES (1992), *Uguali e diversi. Il mondo culturale, le reti di rapporti, i lavori degli immigrati non europei a Torino*, Rosenberg & Sellier, Torino.
King, R. (1987), *Italy*, Harper & Row, London.
Lin, J. (1995), 'Ethnic Places, Postmodernism, and Urban Change in Houston', *The Sociological Quarter*, vol. 36, no. 4, pp. 629-640.
Marra, E. (ed.) (1985), *Per un atlante sociale della città,* Angeli, Milano.
Minardi, E., and Cifiello, S. (1991), *Economie locali e immigrati extracomunitari in Emilia Romagna*, Angeli , Milano.
Montanari, A., and Cortese, A. (1993a), 'Third World Immigrants in Italy', in King R., (ed.), *Mass Migration in Europe. The Legacy and the Future*, Belhaven, London.
Montanari, A., and Cortese, A. (1993b), 'South to North Migration in a Mediterranean Perspective', in King R., (ed.), *Mass Migration in Europe. The Legacy and the Future*, Belhaven, London.
Montani, A. (1966), *La grande migrazione*, Editori Riuniti, Roma.
Natale, M. (1990), 'L'immigrazione straniera in Italia: consistenza, caratteristiche, prospettive', *Polis*, vol. 4, no. 1, pp. 5-40.
Reyneri, E. (1991), 'L'immigrazione extra-comunitaria in Italia: prospettive, caratteristiche, politiche', *Polis*, vol. 5, no. 1, pp. 145-158.
Romagnoli, G. (1995), 'Così è nata la città dei neri', *La Stampa*, 19-11-95.
Romano, M. (1980), *L'Urbanistica in Italia nel periodo dello sviluppo*, Marsilio, Padova.
'San Salvario laboratorio sociale della nuova destra', (1994), special number of *Riff Raff.*
Somma, P. (1991), *Spazio e razzismo*, Angeli, Milano.
Straubhaar, T., and Zimmermann, K.F. (1993), 'Towards a European Migration Policy', *Population Research and Policy Review*, vol. 12, no. 3, pp. 225-242.
Tosi, A. (1993), *Immigrati e senza casa. Problemi, Progetti, Politiche*, Angeli, Milano.
Trufarelli, C. (1989), 'Gli immigrati extracomunitari, con particolare riferimento alla regione Emilia-Romagna', *Parma Economica*, vol. 121, no. 3, pp. 55-61.

5 Urban renewal, ethnicity and social exclusion in France

DANIEL PINSON AND RABIA BEKKAR

Introduction

The study of 'ethnicity' and 'urban renewal' in France differs in important ways from that of the other countries presented in this book. The use of these terms in France grows out of a very specific history of policies and experiences of urban development and immigration. As the comparative studies brought together here will demonstrate, the frameworks for thinking about issues of ethnicity and urbanism in France have evolved out of this particular context, which is quite different in important respects from those studied by our colleagues elsewhere.

To begin with, the term 'ethnicity' poses problems in the French situation. While scholars might use the concept, it has no place in French administrative terminology. Its rejection must be understood in relation to the stated policies of assimilation and territorial dispersal of foreign populations which characterise the French approach to ethnic and national minorities.

In addition, the concept of 'urban renewal' has a particular meaning in French urbanism. At first, it included only the destruction of 19th century buildings and their replacement by new structures in urban areas. But the term took on new meanings from the 1970s onwards, due to the dilapidation and subsequent renovation of large housing projects built in the 1960s to an extent unequalled elsewhere in Europe. Renovation and rehabilitation came to be distinguished.

To address the central questions identified in this book's first chapter, we will look both at renovated neighbourhoods in city centres, and at rehabilitated quarters of urban peripheries. In this context it is possible to understand the position of foreign populations, and particularly the movement that has led a great number of them to abandon the city centre in favour of suburban housing projects.

Institutions, local administration and social exclusion

The very presence of a foreign population in France has produced violent public debates. These polemics have become increasingly bitter during the last two decades in the context of an economic crisis which has created conditions favourable to the development of racist ideas. Nonetheless, as a series of studies have noted, these debates should not lead us to forget the dominant consensus which has been historically constructed in terms of a 'French model', linking citizenship to birthplace rather than to blood (Noiriel, 1988; Weil, 1991; Costa-Lascoux, 1993).

After the tragic epoch of the Vichy government, the rulings of 1945 set up a clear basis for French positions concerning foreigners. These emphasise principles of republicanism and equality for foreigners. These laws formed the basis of the French model until 1993, defining the immigrant as a foreigner who stays in France longer than three months, with no mention of ethnic distinctions (the possibility of including information about ethnicity was initially discussed, but was rejected).

The economic crisis of 1973 marked a change in legal and police controls of immigration as a result of a move from considering immigrants as temporary workers toward seeing them in terms of integration. Following the relative failure of measures designed to encourage immigrant workers to return to their countries of origin, the French authorities moved toward a set of frameworks which permitted employed foreigners to permanently establish themselves and their families in France.

By the end of the 1980s, attempts to revise the 1945 rulings became more pressing. The debates around issues of immigration became more heated in the context of the economic crisis and the increase in racist attitudes, which were reinforced by the rise of certain far Right parties which favoured the exclusion of immigrant populations. These changes must also be seen in light of the Schengen agreements of 14 June 1985, which call for the harmonisation of policies for controlling the entrance of foreigners from outside the European Community.

In 1989, a High Council for integration was set up. The new laws, passed in 1993, must be understood in the context of this commission. These new laws also seek a better control of immigration, a greater efficiency in identity controls, and a less automatic definition of nationality, by which *jus soli* is appended to include a voluntary choice of nationality for children born of foreign parents.[1]

In considering the changes that these laws would introduce, as well as the debates surrounding them, economic realities, human rights issues, and a rallying to the national republican legal tradition must be taken into account (Costa-Lascoux, 1993). In addition, we must remember that the decentralisation laws of 1983 considerably reinforced mayoral powers. In the area of housing, in spite of their inability to influence housing policy and to distribute credits (a prerogative which the central state has maintained for itself) mayors do have the power to accord building permits. They often control the public housing offices (HLM: Habitation à Loyer Modérés), or other building promotion organisations such as the SEM (sociétés d'économie mixte). In this situation, a system of 'passing' immigrant populations to one's neighbours has developed in suburbs of large urban areas .

In general, town officers on the political Right use their local power to oppose setting up programmes for public housing, in favour of more financially viable operations (office space, for example). Left-leaning mayors, although generally more open to HLM enterprises and immigrant families, hide behind designated (but illegal) 'quotas' for foreigners, in order to denounce the egoism of their opponents' communes, and to cope with the pressures that they face because of their unfair practices.

From 1990 onwards, the central state reacted to this situation and took up its role as arbitrator to adopt a series of regulations to control the effects and risks of segregation introduced by such practices. Following the law of 1990, which set up departmental planning for housing under-privileged people, the law of orientation for the city (LOV), adopted on 13 July 1991, establishes the organisation of local housing programmes (PLH: Programme Locaux de l'Habitat) on an inter-city scale to 'assure a balanced and diversified distribution of housing'. These documents condition the state's disbursement of financial aid for a period of three years. However, they must be included in the land use plans (POS: Plan d'Occupation des Sols) which are adopted by the local authorities.[2] The setting up of PLH, which varies considerably according to region, often brings out the contradictions between political rationales and the rationality of urban development.

The history of immigration: spatial distribution

The National Office of Immigration (ONI) was set up in 1945. This government office took the place of private businesses, which had previously run their own campaigns to hire foreign workers. Offices of the ONI were set up in several countries which provided labour, permitting not only a better-organised approach to hiring, but also a better control of those entering French territory. Algerians are not subject to this control.

From 1945-55, entries to France were limited to 32,000 a year: a majority of these were Italians. Then migratory movements accelerated: between 1956 and 1961 more than 500,000 single workers moved to France. Among these, many were Spanish. From 1962-73, the rhythm of entries was even more rapid, with 130,000 more individual workers per year, and with a significant increase in family immigration (60,000 entries per year). During this period these immigrations were still seen as being temporary, and organised ad hoc in terms of measures to 'regulate' the flow.

In 1974, immigrant workers who were permanent residents were 'stabilised', and no further labour immigration visas were issued. The official increase of the immigrant population since then is the result of family regrouping policies (40,000 entries per year). At the same time as traditional labour migration was stopped, measures were taken to encourage the social integration of immigrants in businesses and in society at large (law of 9 January 1973, simplifying naturalisation). After 1981, a policy of integration was developed for immigrants who want to stay in France. It concerned a variety of areas including employment, education, job training, and housing (Noiriel, 1988).

In all, the percentage of the population which was foreign grew as follows:

1946: 4.0 per cent (1,744,000)
1954: 4.1 per cent (1,765,000)
1962: 4.7 per cent (2,170,000)
1968: 5.4 per cent (2,664,000)
1975: 6.5 per cent (3,442,000)
1982: 6.8 per cent (3,680,000)

Urban settings

Today, most immigrants live in urban settings. Seven in ten foreigners live in an urban conglomeration of more than 100,000 people, whereas only about four in ten French do so. Three regions include 60 per cent of foreigners: Ile de France (36 per cent, of which Paris makes up 14 per cent), Rhône-Alpes (12.5 per cent), and Provence-Côte d'Azur. Nationalities are variously distributed throughout the country. In general, Algerians tend to live along a line from Le Havre to Montpellier; Portuguese favour the outskirts of Paris, and its centre; while Tunisians live in Paris and on the south-eastern seaboard. Moroccans tend to reside in Corsica, or to the west of Paris, or in Languedoc; and Italians and Spanish prefer to live near the borders of their respective countries (Villanova and Bekkar, 1994). One foreign family in four lives in public housing (HLM), while only 13 per cent of French do. Indeed, we would argue that access to this government-funded housing, which began in the 1970s, came to exclude immigrants from the advantages of the urban centres (Weil, 1991).

Urban policy

As though in answer to the call of the sociologist Henri Lefevre (1968), it seems that in France, between 1950 and 1990, the right to housing has given way to the 'right to the city'. In France, it is paradoxically the case that it is the most recent sector of its housing, built by the public sector from 1960-75, that is most in need of renovation, due to its physical dilapidation and attendant social problems: housing older than this is generally privately owned.

Following the Second World War, France experienced a spectacular demographic increase: between 1945 and 1970 the urban population progressed at a rate three times that recorded from 1921-46. This exceptional increase brought about a shortage of housing, which the government attempted to palliate by initiating a massive programme of construction. This policy led to the relatively recent nature of the social housing available, while creating a social context favourable to the emergence of problems in poorer quarters.[3] The difficulties experienced by inhabitants of what came to be known as 'quarters of exile' only became fully apparent with the economic crisis and accompanying increases in unemployment (Dubet and Lapeyronnie, 1992).

Housing and public housing projects: a national priority (1950-75)

The housing crisis of the 1950s led to the development of a specific type of housing project.[4] The first projects of 1,000 housing units were constructed in the 1950s. With the development of the ZUP (Zones à Urbaniser en Priorité: decree of 1958) a building boom accelerated: 155 ZUP were created from 1958 to 1964, with a total of 655,000 housing units, covering a surface of 17,000 ha (with an average 4,225 housing units each).[5]

This strong building boom subsequently subsided. A variety of factors contributed: the economic crisis of 1973, and with it the distancing of the state from social projects (many ZUP were left unfinished); but also the progressive redistribution of the housing market between the new HLM housing and rehabilitated housing, in the 'official' framework of the OPAH (Opérations Programmées d'Amélioration de l'Habitat) in old and single family residence sectors (Pinson, 1992).[6]

Today there are 4 million public housing projects, housing 11 million people. They are cheaply constructed, and poorly conceived as to their inhabitability. These physical and social drawbacks lead to their rapid deterioration and abandonment. Such projects are thus an especially unattractive part of the housing market, often more so than the 'old 'quarters or the 'bad' neighbourhoods of the city. An increase in poverty in public housing projects follows their dilapidation, the evolution of which is not without relation to the departure of middling social groups from the public housing to privately owned homes during the 1970s.[7] This took place at the same time as immigrants were moving into the projects, especially in the context of the family regrouping policy of 1974, in spite of the fact that, in order to respect a policy of social 'mixing' in the terms stated by the HLM movement, the social agencies tacitly limited the entrance of foreign families according to a system of 'quotas' set up in collaboration with city officials (Villanova and Bekkar, 1994).

Social urban development (1983) and the 'policy of the city'

The policy of housing followed since 1975 has come to be a part of an overarching programme first called the 'social development of quarters'

(DSQ) and then 'social urban development' (DSU) and now known as 'city policy' (politique de la ville) (Delarue, 1991).

These policies were initiated in 1983, then tested and evaluated over a period of time.[8] They favour local action, and stimulate neighbourhood initiatives. First put into practice in fifteen sites on the outskirts of Paris, Marseilles, and Lyon, this approach came to be associated with other initiatives, particularly ones related to education (called ZEPs; Zones éducatives prioritaires), the fight against crime, and the improvement of transportation linked to job opportunities and training.

Thus, from sixteen initial sites, the number of neighbourhoods covered by these programmes grew to 148 by 1984, then to 300 in 1989. The approach became nearly all-pervasive in quarters considered to be 'difficult'. The programmes are structured by the DIV (Délégation Interministérielle à la ville). At first, they were based on agreements between cities and the state, but they soon came to involve agreements between the central, regional, departmental, and city governments. These contracts and procedures continue today. In November 1994, 214 agreements had been signed with cities, concerning 185 agglomerations and towns, and including about 1,500 quarters (Stébé, 1995). From a policy oriented toward treating urban issues in neighbourhoods (housing, services and public space), there has been a shift towards including economic and social considerations, and involving local leaders, who are legitimated in their action through citizens' expression of their opinions.

Urban renewal in France since 1945

In the immediate post-war period, France had to repair war damage: 425,000 buildings were destroyed during the war, and a million and a half were damaged. What was called 'Reconstruction of France' mobilised the country's financial and business resources until the beginning of the 1950s. Faced with the development of unhealthy quarters in old, overpopulated city centres, the first reaction of the French authorities was to build in outlying areas. The minister responsible for reconstruction and urbanism sought to replace the 'anarchy of housing development' with large, collective housing projects. The hastiness that characterised their building, as well as their later lack of maintenance, rapidly necessitated what was called 'rehabilitation' (as opposed to the 'renovation' of city centres' older buildings).

Modern urbanism, between ZUP and ZRU

The ZUP (Zones à Urbaniser en Priorité) came to be seen as the most efficient solution to the housing crisis, and thus gradually came to take the place of the ZRU (Zones de Rénovation Urbaine) which had been designed to reintegrate insalubrious housing. ZUP constructions had to be underway prior to undertaking demolitions and relocating the inhabitants of the central slums.[9] The vacating of old slum buildings was not accomplished without the passive and active resistance of those occupying them (Coing, 1965; Castells, 1972). The initial goals of this total policy could not be met, and had to be modified. The original goals were to declare the 167 ZUP and their 650,000 initial buildings, as well as the 260 ZRU and their 225,000 buildings.

These new buildings were used to relocate French families who had lived in cramped conditions in the slums, at a time when the need for housing increased with the arrival of the French of Algeria (900,000 arrived in the 1960s) and of two million foreign workers. These workers inhabited old sections of the city, and many set up house in hovels next to the ZUP construction sites where they worked. This situation led the French authorities to intensify the fight against precarious and unhealthy housing. After the laws of 1964 and 1966 against shanty towns, and following the law of 1970 against insalubrious housing, only ten years elapsed before shanty towns essentially disappeared.[10] At the end of 1973 they still housed 60 per cent of the foreign population. On the destruction of their shanty towns , many foreigners moved into situations which were hardly better, particularly into furnished hotels in old quarters of the city. At the same time a series of measures were taken against the 'sleep merchants' (les marchands de sommeil) who exploited the lack of housing by renting overcrowded rooms to foreign workers (Weil, 1991).

From overhaul to renovation

While the ZUP lost momentum in the face of the looming economic crisis, some French families who had lived in housing projects were able to take advantage of savings, made during the previous period of economic growth, to purchase their own homes. In city centres destined to be 'reconstructed' in the style of the ZUP, local movements were set up to protest against this mode of 'renovation' (Castells, 1972). The authorities themselves could not agree on the strategy to adopt in these areas. The

Malraux law of 1962 already served to delay operations which failed to consider the heritage value of old city centres. While in 1972 about fifty 'preserved sectors' were legally designated, and a unanimous rejection of brutal renovation projects became very popular, those in favour of alternative soft or light renovation nonetheless had radically different perspectives on how renovation projects should be carried out. The main conflict arose between those who sought to establish an elitist city centre, and those who wished to maintain populations in existing neighbourhoods (Bourdin, 1984).

Renovation opened up a new approach to the development of old neighbourhoods. The preceding renovation projects systematically demolished old buildings, but by the mid 1970s, rehabilitation projects began to conserve existing structures while refurbishing their interiors. This new perspective was quite influenced by Italian experiments, especially those conducted in Bologna. These projects were complex and often costly, and thus helped to prop up a waning market for the building industry, which had been fundamentally restructured around the massive building projects in the early period.

Starting in 1977, these procedures took on a clearer form with the creation of OPAH (Opérations Programmées d'Amélioration de l'Habitat), which was not really written into building law until the law on city planning (Loi d'Orientation sur la Ville) of 1991.[11] The contractual nature of the process is stated in the memo of 1977, where we can see its relationship to the fragmented nature of property ownership which characterises these older quarters. The approach is fundamentally different from the state policies for the ZUP and ZRU. A partnership is set up for a period of three years, between the city in question, private investors, and the central state, through the intermediary of the ANAH (Agence Nationale pour l'Amélioration de l'Habitat), an organisation created in 1970 to manage funds for investors.

The preparation of the operation is based on initial research, then by studies of the viability of the operation itself. The operation is carried out by specialists and evaluators. In general, these procedures concern small projects which do not include more than 150 units, so large cities often progressively undertook several projects. About 1,800 OPAH were started in France between 1977 and 1989. The relative success of these operations has often had unanticipated consequences: available rental units have been reduced by 70,000 per year. Thus, while the OPAH increased the quality and hygiene of housing, they also reduced the 'market of the unhealthy' (de Rudder, 1978) at a point in time when economic insecurity was increasing; and this contributed to the number of people without housing (400,000 in 1989).[12]

Rehabilitation of social housing from the 1960s and early 1970s

A new procedure was simultaneously set in place for HLM housing in the context of large projects and ZUPs.[13] Called the HVS (Habitat et Vie Sociale), it sought to reinstate the sense of 'liveability' which had been lacking in public housing due to the poor quality of construction, the lack of maintenance, and overcrowding. This led to high charges. The precursor of the DSQ (Développement Social des Quartiers), these measures were essentially technical, and were related to types of relationships with the inhabitants, which revealed other problems.

The lack of respect of the HLM organisations for their renters was quite obvious. Indeed, the development of the office led to an increasingly bureaucratic type of operation. The decentralisation of offices, the creation of local offices in the housing projects, and the participation of residents in decisions concerning the quarters and the management of renting, were often adopted in the framework of the DSQ to promote better communication between all involved parties.

Other aspects of these initiatives included establishing athletic and community associations, making efforts to encourage relations with other neighbourhoods (transportation), the beautification of buildings and their surroundings, the creation of public services and parks, and the introduction of new businesses. These projects sometimes included areas which made them more than simple renovation or rehabilitation projects, but rather examples of 'reurbanisation' (Pinson, 1992).

The consequences of renovation for the exclusion of minorities

Until 1975, shacks at construction sites, furnished hotels in old quarters managed by 'sleep merchants', squalid buildings, and shanty towns, dominated the immigrant housing scene. National measures in favour of the reduction of unhealthy housing (RHI: Résorbtion d'Habitat Insalubre) were at first accompanied by special arrangements for single male immigrant workers, in the form of hostels.[14] Shanty towns progressively disappeared. In the 1968 census, 400,000 people lived in slums, including one in five immigrant families. Some nationalities were more prevalent than others: 9 per cent of Algerians lived in shanty towns, while 32 per cent lived in unhygienic quarters.

The process of moving into normal housing was to be particularly long: nine years was the average stay in precarious housing, and after 16 years of living in France only 6 per cent of foreigners gained access (by 1968) to HLMs. This figure differs according to nationality. While 15 per cent of French lived in HLMs at this time, ten per cent of Italians did, yet only 3.4 per cent of North Africans had gained access to public housing. The 1975 census shows some progress. From 1968 to 1975, the percentage of foreigners living in resident homes went from 9.5 per cent to 6.3 per cent, while those living in temporary housing went from 3.3 to 1.1 per cent.

Since 1974, immigration of single workers has been stopped, and new migrants officially arrive only to become reunited with their families. From this point on, immigration policies have been less oriented toward 'the improvement of reception' and more toward 'the development of integration policies' (Ballain and Jacquier, 1987). A policy for dispersing immigrant families amongst different housing projects has been deliberately chosen. They are asked to live amid French families, in order to participate in social 'mixing', according to the HLM movement's conceptions; and to progressively adopt, if not the French lifestyle (itself a very 'mixed' concept), at least the essential rules of 'neighbourliness' that lead to peaceful interactions. But for this policy to continue to be pertinent, it was necessary that immigrants remain a minority in the HLMs. 'The choice appears logical', comments P. Weil (1991, p. 395): 'between quotas and ghettos, one clearly chooses quotas'.

Another measure taken to better immigrant housing was the use of the 'housing payment' tax, equal to 0.1 per cent of profits, required of all businesses. Part of this tax (25 per cent in 1975) was used for immigrant homes, while the remainder was given to social housing organisations for building. In exchange, a set number of units had to be reserved for immigrant families. Paradoxically, the implementation of this rule, variously interpreted and influenced by the contrasting situation of HLMs, finally resulted in a situation totally different from what had been planned. It accentuated the concentration of immigrants, rather than reducing it (Abou Sada, 1989; Weil, 1991[15]).

It is important to note that this integration policy by a quota system was taking place at the same time as French families were leaving the

HLMs with the aim of buying their own homes. This process was encouraged by the reform of housing finance in 1977. By replacing 'aid for stone' by 'aid to the person',[16] it offered the possibility for families of modest means to obtain financial assistance to buy homes (Aide Personnalisée au Logement); but it also developed a system of unequal rents in housing projects, and a tendency to look to the state to solve each and every problem. Feelings of injustice developed in the midst of a climate of insecurity and economic crisis (Bourdieu, 1993). Only 10 per cent of North Africans own their homes, while 55 per cent of the French do.

Based on 1982 statistics, studies show that foreign households were forced to leave Paris to live in HLM housing in the suburbs (Villanova and Bekkar, 1994). Rehabilitation projects in older quarters had the effect of evicting these populations, due to the rising value of real estate and the consequent rise in rents, which meant that the buildings no longer met the needs of single workers (Abou Sada, 1989; Blanc, 1990; Authier, 1993).

The central authorities, responsible for upholding republican values, were incapable of offering an economic alternative outside the resources of the increasingly decentralised welfare state. They attempted without success to render foreign populations 'invisible', and to ward off racist reactions to them, by dispersing them through the urban landscape. This was the aim in the HLM projects, which, unlike housing in the private sector, were controlled by the authorities.

Case studies

The operations presented below share several characteristics: they are all relatively recent (the beginning of the 1980s), and they were all set up by municipalities in old neighbourhoods dating from the 19th century, with very visible immigrant populations - both in terms of commercial and residential uses. The renewal projects were undertaken in the context of general urban policies (DSQ, then DSU), which indicates that they were conceived in the context of projects which went beyond simply building. They consisted of the destruction of certain sectors, often under projects for the renovation of insalubrious housing (RHI), and the replacing of some of them by social housing (HLM) or by public facilities. Others used procedures designed for the improvement of housing (OPAH: Opérations

Programmées d'Amélioration de l'Habitat), to transform sectors which were privately owned.

Nonetheless, in the different cases, the social effects differed. In the Belzunce quarter of Marseilles, people of foreign origin generally remained in the area and participated in the renovation programme. In Paris's 'Goutte d'Or' and Belleville neighbourhoods, comparable populations tended to leave the city and set up in the suburbs.

The Belzunce quarter of Marseilles

Contrary to popular opinion, the immigrant population of Marseilles is not the highest in France. In 1982, only 6.9 per cent of the population of Marseilles was foreign, a figure which rose to 9 per cent by 1982. These numbers do not, however, tell the whole story, for they do not take into consideration the large percentage of North Africans, and particularly Algerians, who have been established in Marseilles for a very long time. They are particularly visible in the city centre, and contribute to the very identity of the city. A study of the relationship between social and spatial differentiation outside of Paris shows that the quarter of Belzunce in Marseilles displays a type of urbanisation unlike anywhere else in France. The area is populated mainly by single male Maghrebians (Mansuy and Marpsat, 1994).

The distribution of immigrants in the areas around Marseilles From 1950-75, Marseilles went through major economic and political transformations. In the midst of the post-war economic boom, and extremely implicated in the Algerian conflict, the city saw its population increase by 250,000, due to rural exodus, Corsican migration, and Algerian returnees. A large port and industrial city, it was then hit hard by the economic crisis. Between 1975 and 1990, Marseilles lost 100,000 inhabitants, and moved toward becoming an administrative city.

The integration of immigrants that had taken place, both in the work place and the neighbourhood, became problematic in the context of North African emigration, and particularly with respect to Algerians, in the context of the war and its repercussions. A general feeling of mistrust flourished in working class groups, as well as in the police force. This hostile climate had an impact on administrative decisions of authorities concerning Algerian workers.

In the centre, 'micro spaces of quasi-communitarian social life' were observed (Césari, 1995). In addition, spontaneous housing was erected on construction sites for public housing projects in the Northern quarters, creating shanty towns. By the end of the 1960s, social housing programmes were initiated, to produce housing designed for individual workers.

From about 1975, in the context of family reunification policies, some immigrants were able to move into the Northern neighbourhoods of the city. After the progressive departure of middling social groups, and in spite of the application of quotas, by 1990 18 per cent of the inhabitants of HLMs were foreign. Extremely large concentrations of foreigners occurred in some projects. J. Césari writes of 'ethnic specialisation' in respect of these buildings, and even regarding hallways in buildings (Césari, 1995).

Initiatives to recapture the city centre Although the Northern parts of the city were involved in an operation for social urban development from 1983, one of the first experiments of the 'city policy' (DSQ), they also took part in the GPU (Grand Projet Urbain) of Marseilles, which was then presented as the great remedy for urban problems. It involved ten extremely problematic neighbourhoods. In Marseilles the goal was to 'reunite' these problem-filled northern areas with the rest of the city. This initiative was, however, overshadowed by a more ambitious project with which it overlaps: Euro-Mediterranean. This second programme proposes to encourage the port's activities within a framework of Mediterranean regional development, linking port development to land-based economic activities.

Already, in 1977, the city of Marseilles had stated in its plans that it wanted to make its centre 'a public space, open to all'. The Belzunce quarter is at the heart of this old project for the recapture of the centre; since the city cannot disperse its immigrant population, it seeks to render the social context of the neighbourhood 'invisible'. The quarter is in central Marseilles, located near the port. Its buildings are characteristic of the 18th and 19th centuries, and have increasingly not been maintained, and have thus depreciated. These older buildings have been used for immigrants throughout the 20th century, while the upper and middle classes of the city invest in newer buildings in the southern part of the town (Roncayolo, 1981). Today a large proportion of unmarried Maghrebian and African migrant workers live in the upper floors of these buildings, while the ground floors house businesses of 30 to 60 square metres. The proportion of unqualified workers is unequalled in other French areas: it reaches 28.6 per cent. Small shop owners make up 7.1 per cent of the population. While the area was affected by the Algerian war, it has remained an important point of exchange for international and particularly North African goods.

The paradoxes of the renovation of Belzunce The issue of the recapture of Belzunce is complex, for while there is an implicit desire to evict poor families and single workers, this must be accomplished without bringing about the departure of well-to-do shopkeepers. The ZAD (Zone d'Aménagement Différée) has enabled the city to set up a series of projects that range from 1960s type massive renovation (Centre Bourse), to conservative rehabilitation of the historic centre (Vieille Charité).

Between 1981 and 1983 the operation was carried out in co-operation with the population of the area (Mazzella, 1996). A specific division of labour was brought about: the city took charge of public buildings, the North Africans their restaurants, the Jews their fashion boutiques, and the Africans their apartment houses (Tarrius, 1992). In 1983, a local commission was set up. Its task was to complement the activities of the municipality via social initiatives. It contributed to the opening of a baby-sitting centre, a sports and cultural centre, a police station, and a community centre. Two years later, the city set up a 'central city mission'. The aim of this initiative was to develop a more active policy in terms of a 'rationality of openness' and of 'urban mixing'.

Problems soon appeared between the partners of the city and the DSQ. A development commission was therefore set up, to reconcile the desire to rehabilitate with that of maintaining social considerations in the project. Taking into account the movement of the city centre toward other areas, especially near the Estienne d'Orves plaza, the new commission oriented its action towards preserving the traditional functions of the quarter, as well as the integration of new inhabitants. By the end of the fifth and last year of the DSQ contract (1989), the kinds of initiatives were more open, and the aim that the quarter should be a 'hyper centre' was confirmed. In de-stigmatising the area, there was a movement from 'thinking in terms of a quarter to thinking in terms of a centre' (Mazzella, 1996).

In general, North Africans went along with these initiatives and the renovation movement, by putting into practice strategies for the 'sur-valuation of micro-spaces' (Tarrius, 1992). In real estate, for example,

North African buyers, often experienced in commerce, managed to sell property to arriving immigrants for up to five times the going price, in spite of a depressed real estate market. Thus the renovation projects did not in fact encourage the North African or Central African migrants to leave the quarter, nor did they stop the arrival of new migrants, much to the dismay of certain other social groups in Marseilles (Tarrius, 1992).

Two examples of urban renewal in Paris: the 'Goutte d'Or' and Belleville

The effects of urban renewal in Paris have been the subject of many studies. Among those most often cited is Henri Coing's *Urban Renewal and Social Change*, an edited volume published in 1966. That series of studies looked at the Place d'Italie neighbourhood. Here, we will present two more recent case studies, because they concern quarters which have large immigrant populations: the Goutte d'Or (literally 'the Drop of Gold'), and the neighbourhood of Belleville.

The transformation of a pluri-ethnic quarter: the Goutte d'Or The Goutte d'Or is situated in the 18th district of Paris, at the base of the Montmartre hill. The area offers easy access to the metro, the regional transport system, and international train stations. For many decades it has served as a transitional space for foreigners arriving in Paris. Algerian Kabyles have lived in the area since the 1920s, as have Italians, Belgians, Russians, Poles, and many other nationalities. In the 1950s, the area was seen as one of the centres of the Algerians of Paris. Today, it includes twice as many foreigners as other parts of eastern Paris: four in ten people in the neighbourhood are foreign.

The 1970s formed an especially vibrant period in the history of the Goutte d'Or. The numerous hotels, cafes, and textile and jewellery shops owned by North Africans flourished. With the introduction of family regrouping practices, national identities were reinforced. The commercial diversity of the quarter was enhanced. Yet changes also took place as new ambitions, new ideas for the future, and changing perspectives, took hold among immigrants: especially North Africans. Today, Caribbean and African businesses are taking the place of the North Africans. These businesses constitute social as much as economic spaces. They allow for a whole variety of exchanges. They offer housing for people passing through, or hand on information to others, thus answering the needs of groups who are in the process of quickly setting up house.

The Goutte d'Or has been the object of an intense renovation programme, due to this strong presence of immigrants and the number of ethnic businesses. Prior to the renovations, two researchers, J.-C. Toubon and K. Messamah, made an in-depth study of this multi-ethnic space. Their research was published in 1990 and has since become a classic in France. In addition to their very thorough anthropological contributions, which help us to understand the way the quarter functions, their conclusions detail their perspective on the social consequences of the proposed urban renewal projects.

The first projects began in 1983. They included the rehabilitation of unhealthy housing in certain areas (RHI), particularly in the Southern part of the quarter. They also put into practice a system of acquisition and expropriation directed by an organisation controlled by local officials. They then continued in the form of several OPAH designed to help private owners to renovate their buildings. New building included a nursery school, a parking lot for the day care centre, and a police station.

The social transformations brought about by the renewal process were related to the insufficient number of homes provided: 1,400 housing units and 400 rooms were destroyed, whereas only 650 housing units were rebuilt. The types of housing provided were not determined by demand. Although the area has a majority of single people, small apartments are lacking. Similarly, the relocation of migrants who lived in hotels was not taken into consideration. Hostel rooms were allocated to handicapped people over 60 years of age, and to anyone over 65 who had sufficient means. Other single people were told that they should look for housing in the suburbs, in spite of the fact that a poll indicated that 60 per cent of the people questioned wanted to remain in the neighbourhood. People in unstable situations, who had been part of tightly knit social networks, risked becoming disturbed when taken out of this milieu, and some invariably became social outcasts.

In addition, many business people invested their eviction compensation in other neighbourhoods, especially in the light of the incredibly high cost of 'rehabilitating' their locales according to the new norms. J.-C. Toubon and Messamah suggest that the economic space of the quarter will thus be progressively homogenised, as these shopkeepers leave, and are replaced by companies which already have premises on nearby boulevards.

In conclusion, the Goutte d'Or will undoubtedly, little by little, lose its specific character, and most of its inhabitants will be forced out beyond the city limits. No exact estimate has been made of the social consequences of the situation thus far.

Belleville: city government infringement on a working class area For a very long time the Belleville area, situated on a hill in the northern part of Paris, was seen by Parisians rather as the Bronx is today perceived by New Yorkers. The bastion of the 'dangerous' classes, prompt to mobilise during the 19th century, the commune of Belleville became part of Paris only in 1860. After the First World War, this working class neighbourhood began to receive a wave of foreign immigrants, who set up house in its small apartments, which were available in former workers' homes built at the end of the 19th century. Beginning in the 1950s, the percentage of foreigners quickly rose: it was 9 per cent in 1954, 13 per cent in 1962, 20 per cent in 1968, 23 per cent in 1975, and about 25 per cent in 1990. The most prominent nationalities were Tunisian, Algerian and Yugoslavian.

Belleville has been the subject of many studies, which have often been motivated by the rapid changes observable in the area (Simon, 1994; Bekkar, 1997). A first phase of renovation in the area was especially brutal. Between 1956 and 1964, nearly four thousand buildings were destroyed, and replaced by 'towers' of HLM public housing. These transplants clearly stand out. A series of new operations was again initiated at the beginning of the 1980s. A ZAC (Zone d'Aménagement Concertée: procedure which replaced the ZUP) and then a DUP (Déclaration d'Utilité Publique: less complicated for the city to undertake than the ZAC) were begun. Another ZAC is currently being started. Several OPAHs have been implemented to improve privately owned buildings.

The social consequences of these plans are readily apparent. If, at first, relocation of residents could be handled within the neighbourhood, this is now discouraged by 'gentrification'. Social housing characterised by low rents but handled by private owners has become scarce. The furnished hotels which housed older male immigrants were especially hard hit. Publicly owned housing is, in fact, difficult to obtain for those with the lowest salaries: thus only 9 per cent of immigrants were able to obtain flats constructed since 1960 in Belleville. The majority of the foreigners who had lived in Belleville since arriving in Paris have now been relocated in the suburbs, especially at St. Denis and Plaisir. The latest DUP (Déclaration d'Utilité Publique), known as 'Bisson Ramponneau', with its

258 new apartments, replaced 430 apartments and 30 hotel rooms. The difference requires no comment.

In general, urban renewal projects written into law bring about a more rapid dilapidation of buildings than would otherwise be the case. Private owners hesitate to undertake costly maintenance. Real estate agents are less inclined to purchase buildings, and often, if they do buy them, immediately board or brick up the windows to discourage squatters (Simon, 1994). Some parts of Belleville have become veritable urban wastelands, where inhabitants must cross construction sites, and where access to the rest of the city is made difficult by these legally sanctioned obstructions. In some cases everything is done to get people to sell their homes and leave the neighbourhood.

Middle Class people have progressively come to live in areas near those being renovated. The prices of the limited number of commercial and industrial properties have risen. They have become inaccessible to many artisans, who tend to leave the neighbourhood. In the old Belleville quarter, workshops and homes were mixed together. If homes or apartments were purchased, it was not costly to set up a shop in the yard or courtyard. Today, on the other hand, the high costs of freeing lots for construction makes it necessary to require areas not used for housing to shoulder the bulk of property taxes. Rents for work spaces are often ten times higher than in old apartment buildings. Given this situation, many artisans have stopped work to retire, but many others have simply swelled the ranks of the unemployed.

The most recent planning project introduced by city officials has brought residents together to develop an organised expression of their positions. An association, called la Bellevilleuse, has been set up for the defence of the Belleville neighbourhood. Its members have sought counter-expertise with respect to city renewal programmes, and obtained the re-examination of procedures through the courts. Paradoxically, these neighbourhood proceedings are a patent sign of the gentrification of the area, which has become a haven for artists, intellectuals, and managers, seeking to preserve a certain lifestyle. Meanwhile, poorer populations have simply had to leave Belleville.

Directions for future research

Several promising areas of research which remain little explored in France should be mentioned:

- new conceptions of citizenship - particularly European citizenship. (In France, citizenship and nationality are indistinguishable. The civil 'space' is thus reduced and subject to national politics and the affirmation of the centralised nation-state. The English term 'citizenship' has no true equivalent in French. There, the idea of a democratic polity - civil society - outside of national political conflicts has little meaning: see Dubet and Lapeyronnie, 1992);

- taking into account ethnicity in the housing question, and moving beyond taboos born of republicanism, which encourage racist arguments (Wieworka, 1996);

- closer study of cultural practices (communities, associations, commerce, domestic and urban practices), especially in respect of economic exchanges and new forms of collective self help;

- analysis of segregation and integration dynamics, especially as developed by the Housing Observatory;

- the study of population displacements after renovation operations, including access to activities and services.

Notes

1 According to J. Costa-Lascoux, the 'French exception' resides in the large number of ways in which one can be naturalised, in accordance with the idea that social relations are essentially political rather than factual or ethnic.
2 The POS (Plans d'Occupation des Sols) have set the framework for urban planning in towns in France since the land property law of 1967, by determining clearly regulated functions for land use. These documents serve in issuing building and development permits, and can be opposed by a third party.
3 By 1978, after this building boom, half the country's buildings had been erected since 1949.
4 The reappropriation of the 'Athens Charter' approach to urban planning, as adopted in 1934 and elaborated by Le Corbusier during the war, plays a part in this policy, as

do social needs, and the context of massive building projects to meet the housing shortage (Pinson, 1992).

5 The scale of these operations was enormous. The average size was no less than 4,000 housing units.

6 Thus, the production of new buildings receiving state aid diminished by half in twenty years (250,000 in 1993, versus 500,000 in 1972).

7 In 1978, 12 per cent of the population living in HLMs belonged to the part of the population earning salaries in the bottom 25 per cent. By 1994, this figure had reached 36 per cent.

8 Their beginnings are generally seen as a response to the riots that took place in the suburbs in 1981 following the election of François Mitterrand as president.

9 The ZUP of 4,000 at La Courneuve thus came to house Parisians from the Belleville quarter (cf. case studies).

10 The La Courneuve shanty town was a good example of this. In 1966 there were 255 shanty towns in France. They were concentrated in the Paris area, but Marseilles also had 7,800 people inhabiting this type of housing. The 1968 census showed that 400,000 people still lived in slums.

Precarious housing remained the main type of habitat for immigrant workers and families into the 1970s, constituting what Véronique de Rudder has called the 'market of unhealthiness' (de Rudder, 1978).

11 Previously the preserved areas indicated in the Malraux law gave birth to the 'grouped projects for building restoration'. Between 1969 and 1976, 47,000 housing units were thus renovated.

12 It is interesting to note that half of the OPAHs were set up in rural areas.

13 2,800,000 housing units were constructed in this way in post-war France.

14 Two important policies were implemented concerning foreigners at the end of the 1950s, with the creation of the SONACOTRA in 1957, and then the FAS (Fonds d'action sociale) in 1958 (Ballain and Jacquier, 1987).

15 'Contradicting the original project, empty apartments in the most run down of the large housing projects were rented to immigrant families, without consideration of the concentration of foreigners in the buildings .... Thanks to financial incentives, new buildings were built or renovated and offered to other renters.' Weil, P. (1991), *La France et ses étrangers*, Calmann-Lévy: Paris, p. 401.

16 The HLMs were constructed with long term credits with low interest rates, collected by the State and managed by the CDC (Caisse des Dépots et Consignations). Rents were thus 'moderate', and the same for all tenants regardless of their financial situation. It was the product, and not the family earnings, which determined the rental price. With the 1977 personalised housing aid package (Aide Personnalisée au Logement), managed by the family aid services (Caisse des Allocations Familiales), families paid a 'leaving rent' based on their income.

References

Abou Sada, G. (1989), 'L'insertion des immigrés: l'exemple du logement', in *Revue des Affaires sociales*, no. 3.

Authier, J.-Y. (1993), *La vie des lieux, un quartier du vieux Lyon au fil du temps*, PUL, Lyon.

Avenel, C. (1996), 'Quartiers défavourisés et ségrégation', in *Hommes et migrations*, no. 1195, February, Paris, pp. 34-40.

Ballain, R. and Jacquier, C. (1987), *Politique française en faveur des mal logés (1945-1985),* Rapport GETUR, Ministère de l'Equipement, du Logement et de l'Aménagement du Territoire, Paris, pp. 257-283.

Barou, J. (1989), 'L'insertion des immigrés passe par leurs conditions de logement', in *Hommes et migrations*, no. 1118, January, pp. 29-37.

Battegay, A. (1992), 'L'actualité de l'immigration dans les villes françaises: la question des territoires ethniques', in *REMI*, vol. 8, no. 2, Poitiers, pp. 93-98.

Bekkar, R. (1997), *Territoires et publicisation des religions à Belleville. La vulnérabilité des rencontres*, Fonds d'Action Sociale pour les travailleurs immigrés et leurs familles, Paris.

Blanc, M. (1990), 'Du logement insalubre à l'habitat insalubre dévalorisé' in *Les Annales de la recherche urbaine*, no. 49, *Immigrés et autres,* December, MELTM, Paris, pp. 37-47.

Boumaza, N. dir., (1989), *Banlieues, immigration, gestion urbaine*, Institut de Géographie Alpine-Université J. Fourrier, Grenoble.

Boumaza, N. (1992), 'Les relations interethniques dans les nouveaux enjeux urbains', in *REMI*, vol. 8, no. 2, Poitiers, pp.101-119.

Bourdieu, P. dir., (1993), *La misère du monde*, Seuil, Paris.

Bourdin, A. (1984), *Le patrimoine réinventé*, PUF, Paris.

Brun, J. and Rhein, C. (1994), *La ségrégation dans la ville*, L'Harmattan, Paris.

Burgel, G. (1993), *La ville aujourd'hui*, Hachette, Paris.

Castells, M., (1972), *La question urbaine*, François Maspéro, Paris, pp. 377-398.

Césari, J. (1995), *L'islam et la politique en France. Les modalités d'application d'une condition minoritaire. L'exemple de la population maghrébine à Marseilles*, Thèse de doctorat en Sciences politiques, Université d'Aix-Marseilles III.

Coing, H. (1966), *Rénovation urbaine et changement social*, Les éditions ouvrières, Paris.

Costa-Lascoux, J. (1993), 'Continuité ou rupture dans la politique française de l'immigration: les lois de 1993', in *REMI*, vol. 9, no. 3, Poitiers, pp. 233-261.

Delarue, J.-M. (1991), *Banlieues en difficulté, la relégation*, Syros Alternatives, Paris.

Donzelot, J. (1994), *L'Etat animateur*, Editions Esprit, Paris.

Dubet, F. and Lapeyronnie, D. (1992), *Les quartiers d'exil*, Seuil, Paris.

Duby, G. and Roncayolo, M. (1985), *Histoire de la France urbaine. T V: la ville aujourd'hui*, Seuil, Paris.

Duclos, D. (1974-75), Rénovation urbaine et capital monopolistique à Paris, in *Espaces et société*, no. 13-14, pp. 135-143.

Dupaquier, J. dir., (1988), *Histoire de la population française, T IV: de 1914 à nos jours*, PUF, Paris.

Grafmeyer, Y. (1994), 'Regards sociologiques sur la ségrégation', in *La ségrégation dans la ville*, L'Harmattan, Paris, pp. 85-117.

Haumont, N. (ed.), (1996), *La ville: agrégation et segrégation sociale*, L'Harmattan, Paris.

Kaufman, J.C. (1983), *La vie en HLM. Usages et conflits*, Les éditions ouvrières, Paris.

Lebon, A. (1990), *Regards sur l'immigration et la présence étrangère en France*, Rapport DPM, Ministère des Affaires sociales, Paris.

Lebras, H. (1994), *Le sol et le sang*, Editions de l'Aube, La Tour d'Aigues.

Lefevre, Henri (1968), *Le Droit à la Ville*, Anthropos, Paris.

Mansuy, M. and Marpsat, M. (1994), 'La division sociale de l'espace dans les grandes villes françaises hors Ile de France', in Brun and Rhein (eds), *La ségrégation dans la ville*, L'Harmattan, Paris, pp. 195-227.

Mazzella, S. (1996), *L'enracinement urbain: intégration sociale et dynamiques urbaines, les familles maghrébines du centre ville de Marseilles*, Thèse de Doctorat de Sociologie, EHESS.

Noiriel, G., (1988), *Le creuset français Histoire de l'immigration XIX-XXe siècle*, Seuil, Paris.

Pétonnet, C. (1992), *Espaces habités, ethnologie des banlieues*, Galilée, Paris.

Pinson, D. (1992), *Des banlieues et des villes, dérive et eurocompétition*, Les éditions ouvrières, Paris.

Pinson, D. (1995), 'Générations immigrées et modes d'habiter, entre repli communautaire et fusion transethnique', in *Les Annales de la Recherche urbaine*, no. 68-69, *Politiques de la ville, recherches de terrains*, Plan Urbain, MELTT, Paris, pp. 189-198.

Rodriguez dos Santos, J. and Marié, M. (1973), 'L'immigration et la ville', in *Espaces et sociétés*, no. 8, February, pp. 223-226.

Roncayolo, M. (1981), *Croissance et division sociale de l'espace. Essai sur la genèse des structures urbaines de Marseilles*, Thèse de Doctorat d'Etat, Université de Paris I.

Rudder (de), V. (1978), *Le marché de l'insalubre, in Espaces et société*, no. 24-27.

Schnapper, D. (1992), *L'Europe des immigrés*, F. Bourin, Paris.

Simon, P. (1994), *La société partagée, relations interethniques et interclasses dans un quartier en rénovation: Belleville*, Thèse de Doctorat, EHESS, Paris.

Stébé, J.-M. (1995), *La réhabilitation de l'habitat social en France*, PUF, Paris.

Taffin, C. (1991), 'Le logement des étrangers en France', in *Economie et statistique*, no. 242, April, INSEE, Paris, pp. 63-67.

Tarrius, A. (1992), *Les fourmis d'Europe: migrants riches, migrants pauvres et nouvelles villes internationales*, L'Harmattan, Paris.

Tarrius, A. and Péraldi, M. coord., (1995), 'Marseilles et ses étrangers', in *REMI*, vol. 11, no. 1, pp. 5-132.

Toubon, J.-C. Messamah K. (1990), *Centralité immigrée: le quartier de la Goutte d'Or*, L'Harmattan-CIEMI, Paris.

Van de Walle, I. (1991), *Belleville, un village dans la ville*, Rapport au FAS, Paris.

Verbunt, G. (1996), 'Identités et communautarismes, vrais débats et fausses réponses', in *Hommes et migrations*, no. 1195, February, Paris, pp. 22-26.

Vieillard-Baron, H. (1994), 'Des banlieues aux ethnies, géographie à voir, historie à suivre', in *Les Annales de la Recherche urbaine*, no. 64, *Parcours et positions*, Plan Urbain, MELTT, Paris, pp. 96-102.

Villanova (de), R. and Bekkar, R. (1994), *Immigration et espaces habités*, L'Harmattan-CIEMI, Paris.

Weil, P. (1991), *La France et ses étrangers*, Calmann-Lévy, Paris.

Wieworka, M. (1996), 'Racisme, racialisation et ethnisation en France', in *Hommes et migrations*, no. 1195, February, Paris, pp. 27-33.

Withol de Wenden, C. (1996), *La nouvelle citoyenneté*, in *Hommes et migrations*, no. 1196, March, Paris, pp. 14-16.

6 Local government, ethnicity and social exclusion in Portugal

CARLOS NUNES SILVA

Introduction

This chapter presents a preliminary account of local government's approach to ethnicity[1] and related social exclusion in Portugal. The growth in the number of immigrants is intimately linked to global economic restructuring, and to the new position of Portugal in the international division of labour. Recent immigration to Portugal is associated with ethnic minorities, mainly from the former Portuguese African colonies after independence in 1975. Urbanisation has separated people spatially along economic and social lines, producing significant variations of income and class within cities.

In this chapter we argue that, in spite of ethnicity being of low salience in Portuguese society, local councils in urban areas - mainly in municipalities with a significant concentration of African population - are incorporating new policy instruments directed to ethnic minorities in their main activity plans, year after year. There are also signs that ethnicity is becoming a higher profile issue, in part because of the actions of the local state and those of immigrants' associations. Local government policies have not been deliberately exclusionary, although for economic reasons they tend to separate lower-income housing from that of the more affluent areas. This partly explains the relative concentration of social housing, and therefore of ethnic minorities, in certain sectors of the metropolitan area. There is anecdotal evidence of racial prejudice regarding the location of housing estates with a high proportion of ethnic minority groups (mainly Africans and gypsies).

Local government and local governance in Portugal

There are 305 municipalities and 4,221 parishes in Portugal, with a total population of 9.8 million inhabitants. Local government is responsible for 7 per cent of total public expenditure, which makes Portugal one of the most centralised countries in Europe. Municipalities are responsible for 95 per cent of all local government finance, and parishes for only 5 per cent (Silva, 1996a). The importance of parishes is increasing, and a recent Law (nº 23/97, 2-7-97) enlarged not only their competencies but also their financial capacity. Local government structure has been stable since the 1976 Constitution, with two boards - one deliberative, and the other executive. Both are directly elected in the case of municipalities; in the case of the parish only the deliberative board is directly elected (Silva et al., 1994).

Only nationals could vote and be elected in national and local elections. Partly as a result of the Maastricht Treaty, the situation changed from the December 1997 elections onwards: foreign nationals with permanent residentship became able to vote and to be elected. However, the situation before December 1997 was responsible for the divorce of immigrants from the local political process in Portugal. The only way ethnic minorities could influence local decision making was through immigrants' associations or neighbours' associations, some of them very active in demanding better living conditions. There are, indeed, very few local councillors from ethnic minorities but with Portuguese nationality, and even fewer in senior positions. As we shall see, their small numbers, low degree of social integration, and geographical dispersal - except in some municipalities in the Lisbon Metropolitan Area (LMA) - are reasons why ethnic minorities have so far not been able to organise themselves as a unified bloc in local politics, and have not felt the need to do so.

In formal terms, local government in Portugal has a strong political and administrative position, guaranteed by the democratic Constitution of 1976 and subsequent revisions. It has the monopoly over a range of sectors like urban planning and basic infrastructure, but only a small range of formal responsibilities over local social welfare services (Silva, 1993a; 1995; 1996a; Silva et al., 1994). In reality, however, national government can, and does, exercise control over local governance through legislation and regulations of all kinds (Silva, 1995; 1996b; Silva et al., 1994).

As local government duties are defined by the principle of generality, which means that they can do anything for the well being of the inhabitants of their areas, there is room for differentiation among municipalities. However, there are areas of policy where local governance is mandatory. If it is true that local government makes its own decisions in most areas - urban planning, infrastructure, urban services, cultural activities, etc. - there are other areas where legislation and statutory requirements regulate local government activities. In respect of immigrants and ethnic minorities, there are no specific mandatory requirements on local government. This has been the domain of central government, which over the years has launched several programmes and provided grants for local government to promote social integration, in some cases under EU Initiatives such as URBAN, Horizon or Inclusion.

Immigrants and the ethnic minority population in Portugal[2]

It is difficult to say how many foreigners live in Portugal, and from what ethnic groups, as the number of *illegal* immigrants is assumed to be very large, despite a legalisation process carried out since 1992 (Law nº 212/92 - Law on extraordinary legalisation of immigrants; Law nº 17/96, 24-5-96.) About 39,000 illegal immigrants, mainly Africans, regularised their situation under this extraordinary process. At the end of 1996, the official number of foreigners living in Portugal and fully authorised was 172,912: 1.7 per cent of the total population (INE, 1996, based on SEF data).

Portugal has been, for centuries, a country of emigration, but has been also a country of immigration, for example by slaves brought to supplement the work force, since the 15th century (Tinhorão, 1988). But immigration on a large scale to Portugal is of very recent date: it has been mainly a country of emigration. Official figures suggest that foreign-born persons in Portugal in 1960 numbered 29,579, and in 1981 only 31,627. Immigration from the third world, mainly from former Portuguese colonies, became important only after de-colonisation in 1975. The first significant influx took place in the 1960s, from Cape Verde. Apart from the 1974-75 influx of refugees, which was easily assimilated in Portuguese society, a constant flow of immigrants started then, with peaks in periods of rapid economic growth, and with a clear pattern of immigration of a low skilled workforce.

According to the Department of Foreigners and Borders (SEF), of those 172,912 foreigners living legally in Portugal at the end of 1996 (Table 6.1), 81,176 are Africans (46.9 per cent), and these are mainly (95 per cent, 77,114) from the former Portuguese colonies. Including the Brazilian group, foreigners of Portuguese mother tongue represent more than half (56.2 per cent: 97,196).

Among the 77,114 legal immigrants (by 1996) from PALOP countries (former Portuguese African colonies), the largest group is from Cape Verde (39,546, 51.3 per cent), followed by Angola (16,282, 21 per cent), Guinea-Bissau (12,639, 16.4 per cent), Mozambique (4,413, 5.7 per cent), and São Tomé (4,234, 5.5 per cent). The relative proportion from each country has remained the same in recent years, as has the PALOP share of immigration as a whole (Table 6.2).

However, the census of DEDIAP/CEPAC (1995) enumerates a total number of 80,000 from Cape Verde. This is more than double the number officially registered as immigrants. A similar kind of estimate is made for the other PALOP migrants, which confirms a general belief (e.g. SEIES, 1995) that the real number of African immigrants with foreign nationality living in Portugal is around 140,000 or more. African immigration associations estimate a total number of 300,000 (quoted in Machado, 1992).

Other foreign residents have come from more developed countries, and therefore are in a different position, both economically and socially. Those from European countries represent the largest group: 25.3 per cent in 1996; followed by citizens from North American countries (6.2 per cent). The Brazilian community, at 20,082, is a little smaller than the Cape Verdean one. The rest are very small communities, dispersed, and associated with embassies or multinational companies.

Immigrants from PALOP countries are, in general, younger than the Europeans, with around 50 per cent between 15 and 24 years old, while those from other European countries include a significant number over 50 years of age. Regarding professional status, around 30 per cent of European immigrants work in technical, scientific and administrative jobs, while more than 40 per cent of African immigrants work in production or building construction (CGTP, 1993). According to the evidence collected by the Trade Unions Confederation (CGTP), illegal African immigrants are forced to accept very unfair and hard working conditions because they are constantly at risk of expulsion. A similar social and professional composition is expressed in the 1996 SEF data.

Around 95 per cent of the PALOP immigrants are between 15 and 60 years old, and only 54 per cent have regular work. The great majority of these (83 per cent) work in the building industry and public construction,

7 per cent in commerce and restaurants, 4 per cent in cleaning activities, and 2.7 per cent as technicians and professionals.

Immigrants are mostly concentrated in the Lisbon Metropolitan Area and in the Algarve, the tourist resort in the south. For example, 80 per cent of all immigrants from Africa live in the Lisbon Metropolitan Area.[3] They live mainly in slums or in municipal social housing, the latter being the case in the new rehousing programme specially designed to eliminate slums in the two metropolitan areas (PER). While they now live in ghettoes, in the rehousing process there is a clear aim of mixing them with nationals, in a process not always without difficulties, misunderstandings, and press speculation. While African immigrants are mainly concentrated in the metropolitan areas, Brazilians have a strong presence in the district of Aveiro, while the British and Germans are concentrated in the Algarve.

The DEDIAP/CEPAC (1995) Census, of 106 slums, degraded areas, and social housing neighbourhoods, stated that there were 66,513 people of African origin in 13 LMA municipalities. These figures did not include some small groups in the other 5 LMA municipalities, nor Africans living in private flats dispersed throughout the urban area, nor those in illegal small hotels. Nor did they include those living in containers and barracks inside the areas of the major civil construction works in the LMA, where a significant number of the young and active African population live illegally. The northern part had 48,097 (72 per cent) of all the LMA's African immigrants, compared with 28 per cent in the south (18,416). Amadora, with 15,399, had the largest concentration on the north bank, and Moita, with 6,030, had the largest on the south bank.

The DEDIAP/CEPAC (1995) census claimed that the largest group was from Cape Verde with 40,904 (61 per cent), followed by immigrants from Angola (12,309: 18.5 per cent), Guinea-Bissau (7,423: 11.2 per cent), São Tomé (3,231: 4.8 per cent) and Mozambique (2,440: 3.7 per cent). It is difficult to calculate the exact number of *illegal* African immigrants, but it seems that they are mainly living on the south bank of the metropolitan area. Most of them are subject to exploitation and discrimination.

The 1991 census stated that around 3 per cent of Portugal's population (347,233) was born outside Portugal, 67 per cent of them in Africa. If we add to these around 500,000 *retornados* (those born in Portugal who returned in 1974-75 due to the independence of former Portuguese colonies), and also hundreds of thousands of people who served in the military, we appreciate the strength of the link with Africa and its traditions. The integration and acceptance of these immigrants should therefore be easier than in other European countries.

According to the DEDIAP/CEPAC (1995) study, first generation immigrants tend to live in barracks, some of which function as hotels, owned by one, who subsequently rents rooms to companions. The second generation, living as families, tend to be in areas of illegal housing, but in conditions which, though poor, are better than those of the first generation. Those living in social housing estates are, in general, third generation immigrants. They have children born in Portugal, and are therefore better integrated, and are usually rehoused in a process of urban renewal. They have better living conditions, and fewer social problems.

The DEDIAP/CEPAC (1995) demographic census lists more than 20 settlements with more than 1,000 Africans, and 9 with more than 2,000. The two biggest concentrations are Vale da Amoreira, in the municipality of Moita (south bank), with 4,800 Africans; and Cova da Moura, with 3,170, in the municipality of Amadora (north bank). Some sort of local political mechanism should be introduced in order to allow these people a voice in the affairs concerning them.

The SEIES (1995) research found that in some of the settlements of barracks there is a Community Centre, the main objective of which is to improve the living conditions of the inhabitants, irrespective of their national origin. In most cases these centres were started by a religious group, and in some other cases on the initiative of a group of friends or residents, with the support of local government, the Social Security department, or local schools. Some of these centres have managed to develop projects against poverty. The SEIES (1995) study found 74 immigrants' associations, all of them very recent, over half created after 1991. Some of them are not yet formally constituted.

The empirical evidence collected by Machado (1992), on social and cultural contrasts[4] between the immigrant community from PALOP in Portugal and the Portuguese population, reveals differences between African communities, by countries, but none of them in a confrontational position in relation to the Portuguese population - as seems to be the case with Arab immigrants in France, for example. One possible explanation, apart from the traditional link with Africa, and the fact that hundreds of thousands of Portuguese were born there, is the Portuguese social structure - and its social inequalities - which places a large proportion of Portuguese society at a similar level to that of most African immigrants (Almeida, 1992; Rodrigues, 1989; Quedas, 1994).

The evolution of ethnicity in Portugal will depend on the growth of immigration, the integration of the second generation, and, finally, on the evolution of Portuguese society. Paradoxically, the development of state and local government policies towards ethnic minorities seems to be increasing the relevance of ethnicity in Portugal, as more and more it appears in (local) political discourse and practice.

Exclusion from local governance

It is difficult to say that there has been a national immigration policy. Nevertheless there are some policy instruments marked by a clear desire to assimilate immigrants into Portuguese society, whilst accepting cultural differences, and combating all forms of racism.[5] In this respect, local government has been playing an important role in the last few years, in terms of both social integration and the preservation of cultural identity.

Trade Unions (TUs) play an important role in the defence of immigrant labour rights, but affect only a small number, as most immigrants are in the country illegally. Both TU federations - CGTP and UGT - regularly organise information sessions, and put forward specific proposals on these matters. Nevertheless, the low level of TU membership by immigrants, and the large number of illegal immigrants, reduces the impact of these actions. Social organisations, like Misericórdias and Cáritas, with a Catholic background, also play an important role in the defence of immigrants' rights, as well as in the delivering of social services to them.

According to SEIES (1995), based on information given by the national Department of Employment (IEFP), there is a low number of immigrants applying for unemployment subsidy because of the two main conditions which must be met - to be a legal immigrant, and to have worked for at least six months in Portugal with social discounts. Similar conditions apply to training courses: probably an additional cause of difficulties for the social integration of immigrants.

But there are some cases of openly expressed racism by minority political groups. Some extreme cases, relating to the gypsy population in several parts of the country, confirmed hostility to the non-white population among some sectors of the population (a behaviour one could classify as 'not on our doorstep'). But there is no clear 'sector' of the population being anti-immigrant and anti-gypsy, as there is no organised political party with a racist policy. Urban riots, like those in France, Belgium and the UK, for instance, have no parallel in Portugal.

Although anti-immigration measures are not definitively on the political agenda in a country of emigrants, as a member of the EU and a subscriber to the Schengen agreements requiring the harmonisation of policies for controlling the entry of non-EU nationals, Portugal is forced to implement a restrictive immigration policy. Some of the measures have, on several occasions, been criticised by the political parties in opposition, and by human rights non-governmental organisations (NGOs), after some dramatic cases in airports or harbours, mainly involving people travelling from Africa or South America (Cunha et al., 1996). One consequence of Schengen was the need to legalise immigrants. Policies for this appear to have had limited success in early years, probably partly because of some doubts about the seriousness of the process, in a period of increasingly prominent racist discourse in the press against 'gangs' of young blacks. Yet unlike in several countries, immigrants in Portugal do not have a high rate of unemployment, and therefore do not constitute a heavy burden on the social welfare system. Public participation is low for all social groups, including ethnic minorities; but ethnic divisions become prominent when for some reason there is a concentration of people of some ethnic group, usually related to slums and rehousing. But when that happens, events are similar to those affecting white population in equivalent situations.

Till recently, foreigners could not participate in local elections. In December 1997, for the first time, they were able to elect and be elected under certain conditions, and only in cases of reciprocity between the two countries. For nationals from PALOP (former Portuguese African colonies) and Brazil, it is necessary to have a legal residence in the country for the last two years to be an elector, and four years to be elected. EU nationals do not have this time limitation. Other foreigners need three and five years legal residence respectively (Law 50/96, 4-9-96).

Urban policy and social welfare

The local welfare state

The Portuguese welfare state is a recent creation, and much smaller than the others in the EU.[6] Social housing represents less than 5 per cent of the housing stock, and was not intended to be for immigrants, although recent programmes do include them. The provision of social welfare services (child care, old-age care, health care) is comprehensive, although not because of pressure by immigrants' groups. In education, there are special schemes designed for the itinerant population of gypsies. As the formal mother tongue for most of the immigrant population - those from the former colonies - is the Portuguese language, teaching in the mother tongue is not a major problem, a contrast with other European countries.

The development of extra-curricular activities related to these different cultural backgrounds is already a common practice in schools located near the major concentrations of black people. There are no formal discriminatory criteria, but because of their geographical concentration, children in the black population tend to go to schools located near their ghettoes, therefore reducing the opportunities of intermingling with Portuguese children.

There is no detailed information on unemployment among immigrant women, and therefore it is difficult to know how many stay at home, and how many immigrant children attend day-care centres: information important for understanding their later performance in school. The direct knowledge we have of several situations indicates a very low proportion. In some neighbourhoods with a high concentration of immigrants, local associations deliver these kinds of services. An example is Moinho da Juventude, in the illegal housing estate of Cova da Moura, in the municipality of Amadora. (See Simões et al., 1992, and Capucha, 1990; and see Franca et al., 1992, for a broad view of the Cape Verde community).

Housing demand is still very high, reaching around 300,000 units (or 500,000 in other estimates) for the country as a whole. The effort in social housing construction is far from enough to meet such numbers, in spite of recent progress in this field. It is no surprise, then, to see immigrant ethnic minorities being left at the end of the queue. They do have some priority when living in an area due to be demolished and rehoused, as in the metropolitan areas with the PER programme (slum clearance and rehousing). When this happens, municipalities have a particular concern for social integration. The process is not only one of giving a house but also one of social integration and promotion. But problems of social segregation and of ghetto formation have developed in some of these new social housing estates, especially in the earliest ones, which have all sorts of social problems: poor attendance at school, unemployment, drugtaking, vandalism, etc., in spite of several attempts by the municipalities to improve living conditions.

The creation of RMU - Minimum Guaranteed Income, and the complementary Social Integration Programme - for all families living legally in the country and facing certain deprivation conditions, launched on an experimental basis in 1996 by the socialist government elected in October 1995, has been fully implemented since July 1997, with the involvement of municipalities and parishes in the so called Local Action Commissions (CLA) (Law nº 19-A/96, 26-6-96). There is a concentration of applications in the major urban areas, but statistical data with ethnic details are not published, and therefore it is not possible to reach any conclusions about the extent of its application to ethnic minority families.

The illegal situation of thousands of immigrants has consequences on the school results of their children. In general, schools are not prepared for the integration of different cultures, and associated language difficulties. Access to health care is also limited for these illegal immigrants (Simões, et al., 1992; França et al., 1992).

Unlike in some European countries, local government in Portugal has very limited compulsory obligations regarding social policy for immigrants and ethnic minorities. Nevertheless, there is a large set of diverse actions and initiatives developed by municipalities and parishes towards this population group, especially in the last decade. The empirical evidence suggests a more developed and diversified set of initiatives in urban municipalities than in more rural municipalities in the LMA periphery, related to the number of immigrants living in the area.

Local government social policies for ethnic minorities and immigrants in the Lisbon Metropolitan Area (LMA)

The Lisbon Metropolitan Area (LMA) comprises 18 municipalities: 9 on each side of the river, divided into 210 parishes in 1991. 25.7 per cent of the national population lives there (LMA had 2,535,669 inhabitants in 1991). The communist party has a majority in 8 municipalities (Alcochete, Almada, Barreiro, Moita, Montijo, Palmela, Seixal, Sesimbra) out of 9 in the LMA south bank and in three in the north bank (Amadora, Loures and Vila Franca de Xira). The Socialist party runs four municipalities: three in the north (Cascais, Sintra and Azambuja), and Setúbal in the south bank. The municipality of Lisbon is governed by a coalition between the Socialist party (which provides the mayor) and the Communist party. The

Social-Democrat party runs two municipalities on the north bank (Oeiras and Mafra). The Junta Metropolitana de Lisboa (metropolitan 'government') was set up in 1991, an association of the 18 metropolitan municipalities, chaired by a communist mayor (Silva, 1993b). Its main function is co-ordination of municipal policy positions, in particular towards central government. It is also in charge of the management of some EU structural funds allocated in the Community Support Framework-II for municipalities.

In 13 of the 18 LMA municipalities there are significant concentrations of African immigrants. All local government authorities have actions directed towards these populations. Such actions usually include research on specific characteristics of these immigrant communities (for example, their professional and social situation and housing conditions); subsidies and logistical support to immigrants' associations; rehousing policies, especially under the programme PER; organisation of information meetings, music festivals, and other cultural activities; involvement in partnerships with other agencies in projects for the elimination of poverty; and the creation of a 'municipal council of immigrant communities and ethnic minorities', with consultative powers in the formulation of municipal policies towards immigrants' social integration (in two municipalities, Lisbon and Amadora), and including immigrants' associations.

The municipality of Amadora, on the north bank, has the largest community of African immigrants in the LMA, and in Portugal as a whole. It is a member of the European Network of Cities with Ethnic Minorities, and is twinned with the municipality of Tarrafal in Cape Verde. It has a 'Programme on Ethnic Communities' which has already carried out several activities and projects, including a 1995 seminar on minorities. The municipality also has a project for the promotion of health in the immigrant community. This project includes courses, and the training of local representatives, as well as the dissemination of information. Amadora is also involved in the PER programme, with 5,419 new dwellings. Environmentally degraded areas and slums are mainly occupied by African immigrants.

The six main projects carried out by the municipality in 1995/96 include a variety of activities: creation of a database and a documentation centre specialising in ethnic minorities; development of actions against racism and xenophobia, in partnership with research centres; support to local immigrants' associations and to special commemorations; creation of a Municipal Council for the Ethnic Minorities and Immigrants in the municipality; and participation in programmes directed to women in ethnic minorities (as part of the ELAINE Initiative: a European network of cities with immigrants). The municipality supported the candidature for EU structural funds of a project, 'Cities against racism', promoted by the Anti-Racist Front.

The municipality of Moita, on the south bank, has the LMA's second largest African community, and there is a significant concentration of African population in one large social housing estate, with problems of social integration. The municipality created a special unit - Gabinete do Vale da Amoreira - and an Urban Revitalisation Operation for this area, with a total investment of 734 million escudos between 1996 and 1999, with financial support from European Economic Space mechanisms. The municipality also develops, for example, special cultural programmes in this neighbourhood, with special emphasis on African culture (for example, 'African Week'). There has also been a practice of introducing new teachers to the realities of the municipality through specific actions (reception for new teachers) and the promotion of expositions on these themes, like the Oikos-UNICEF Children of all colours expo in 1995. The municipality is also involved in the PER programme.

In the municipality of Seixal, a number of neighbourhoods have concentrations of several social problems: truancy from school, annual maternity, low level of family involvement in school matters, large families and de-structured families, ethnic conflicts, and precarious employment, among others. These neighbourhoods, many of them slums, are mainly populated by African immigrants and gypsies. They include Quinta da Princesa, Santa Marta de Corroios (68.5 per cent Africans), Quinta do Cabral, and Quinta da Boa-Hora. Within them, the municipality is trying to overcome the problems by activities which include health education, socio-cultural animation activities, inquiries on social and cultural needs, occupational programmes for children and youths, a photograph competition under the subject 'All different, All equals', public urban space improvements, the creation of community centres (in part financed by the EU structural funds under the Inclusion Initiative), the creation of employment centres in partnership with immigrants' associations (e.g. Associação Cabo-Verdeana do Seixal), multi-cultural community programmes, the promotion of residents associations, and the rehousing (under the PER programme) of those living in slums. In recent years, the municipality has developed projects of twinning with the municipality of Boavista in Cape Verde, in order to strengthen the integration of immigrants from Cape Verde with the local community. The guiding principles in this policy area are the priority to communities at risk of social exclusion, the promotion of partnerships with several institutions acting in this field, the encouragement of local immigrant community intervention, and the combating of all forms of racism, in order to construct a new citizenship.

In the municipality of Almada, on the south bank of LMA, social integration policy instruments are directed to those elements of the community at risk of social exclusion. Examples of this approach are the Multicultural Education Programme involving around 300 students in 1995; the rehousing programme, PER, adopted in 1994 and involving 2,156 new dwellings in a partnership between the municipality and two departments of Central Government (IGAPHE and INH); as well as intervention under the Inclusion Initiative, involving 112 families in 1995, in partnership with other agencies; and special programmes of social integration in municipal housing estates (involving cultural animation, inquiries on youth and adults' needs, creation of leisure spaces, and the improvement of public spaces). The rehousing of 179 families living in the Asilo 28 Maio in Porto Brandão is an example of extreme social exclusion that affects people of different ethnic origins in the same way. The participation in the project Luta contra a pobreza is another example of this broad approach to social exclusion independent of ethnic characteristics (involving traditional African folk groups in cultural activities as well as Portuguese folk groups, for instance).

The municipality of Loures has twinning agreements with Maio (Cape Verde) and Matola (Mozambique) for co-operation in policy fields such as education, health, administration, and economic development. It is a member of ELAINE. The municipality also supports immigrants' associations located in the area. The guiding principle is to develop self-responsibility on the part of these organisations. Therefore the municipality gives support to associations, provides incentives for their participation in municipal initiatives, and involves these associations as much as possible in partnerships with other bodies acting in the area of social policy. Examples are, for instance, the support given to the 'Association of People from Guinea', in order to participate in the '1st Children's Inter-Ethnic Festival', support to 'ACRA - Angola's Cultural and Recreational Association', to the 'Mozambique Association', and to the 'Islamic Community', among others. Meetings between different communities are also encouraged and supported. In the municipal main Festival (yearly) there is a special place for the immigrant community; this is occupied by immigrants' associations, and has been considered a success by the municipality. In 1996 one of these associations, Association RegriJovem, was involved as a partner in a large project against poverty. There has also been support for intercultural associations located outside the municipality, since their initiatives have some impact in the municipal area. These initiatives include information campaigns on human rights, north-south co-operation, and information campaigns against discrimination and social exclusion. It is already a common practice to give support to NGOs acting in this field (e.g. CPPC and its solidarity campaign with Angola's children; material support to the 'Portuguese Council for Refugees', for the organisation of a Congress; and support to the 'Anti-Racist Front'). Through the PER programme and other schemes, the municipality has been involved in rehousing people (including members of ethnic minorities) living in barracks and other types of poor housing. These actions are usually complemented by social work in order to achieve better social integration. In one of the worse areas - Quinta do Mocho - the municipality has a special project, 'Integrated Operation for Local Development of Quinta do Mocho'. Health promotion has also been undertaken for the immigrant community.

The municipality of Barreiro, on the LMA south bank, has promoted cultural initiatives having as their main subject Africa (e.g. the Video Festival on Africa: Anthropological Exposition on Africa; and music festivals including music from Africa as well as Portuguese folk music). It is also involved in the PER programme and the Inclusion Initiative. In Sintra, the municipality has also put forward a candidature for the project 'The Anti-Racist Cities', and will create an Observatory to monitor social changes and problems in the municipality. It is also involved in the PER programme, which includes also ethnic minority groups, as well as in the Inclusion Initiative.

But in municipalities like Mafra, on the rural periphery of the LMA, there is no need for special action towards ethnic minorities, as numbers are small. This is reflected in the lack of any special reference in the municipality's activity report in 1994-96, the last three years for which records were available at the time of writing. A similar conclusion can be reached in relation to Palmela and Alcochete, both in the southern periphery of LMA. In Azambuja, another peripheral municipality on the LMA north bank, with a small number of immigrants, the policy profile is similar. This municipality is also involved in the PER programme. There is a twinning agreement with Mosteiros, in Cape Verde, and a solidarity campaign was launched with success on the occasion of a volcanic eruption there, in 1995, suggesting the existence of a high level of empathy with other ethnic groups.

Urban renewal and social exclusion

In EU countries, social exclusion is a growing problem in most urban areas, especially in those affected by industrial restructuring and the associated high unemployment rates, income inequalities, education inequalities, poor housing, low levels of social equipment in certain sectors, growing criminality, etc. In most cities these kinds of development happen in areas with a high concentration of immigrants and ethnic minorities, with difficulties of social integration (CE, 1995; Imrie and Thomas, 1993; 1995; Harrison and Law, 1997; McGregor and McConnachie, 1995).

Compared with the rest of Europe, Portugal is in a different situation, as ethnic minority groups are smaller, in absolute and in relative terms. The consequence of economic restructuring, although it has not assumed as severe a dimension as in many other countries, has indeed hit certain parts of the metropolitan area very hard. But this has not affected ethnic minorities as such, but certain professional groups independent of race.

There are three different types of areas subject to urban 'renewal': old centres, slums, and illegal housing estates. In the first case, the approach has been to reject the functionalist principles of urban renewal, and put the emphasis on urban rehabilitation and revitalisation, which means, among other things, that efforts are directed towards keeping people in the neighbourhood after the conclusion of the 'renewal' work. In most of these areas there is only a minor presence of immigrants and therefore there are no specific problems. In the case of slums, there is indeed a significant proportion of immigrants and ethnic minority groups. The rehousing programmes also include them, which is not the case with other social housing programmes, for which only nationals are eligible. Most of the new houses are built in areas surrounding slums, as part of a more comprehensive urban plan which also includes, usually, housing for other social groups. But there are also examples of rehousing far away from the previous settlement. In the case of illegal housing estates, these were mainly built by the white Portuguese population, and therefore there are no specific ethnic issues in the process of the urban rehabilitation of these areas. In any case, we need to consider the way that urban regeneration impacts on ethnic minority groups in different ways, thus dividing them from each other as well as from the rest of the population.

Since the integration of Portugal in EEC/EU, there has been a strong effort to clear slums from the main urban areas, in particular within the LMA. Events such as Lisbon being selected as European Cultural Capital in 1994, and Expo98, were also responsible for the acceleration of the slum clearance process. The URBAN initiative is being applied in slum clearance and urban rehabilitation in a limited number of areas in the LMA, as elsewhere in the country. The programme's main objective is to fight the major causes of social exclusion, by promoting the creation of business and the improvement of infrastructure and public services, and by offering new opportunities for less privileged social groups (DGDR, 1996). Therefore, in spite of the ethnic mixture, there is a tendency towards the spatial concentration of ethnic groups in certain areas, as a consequence of slum clearance initiatives. The above-mentioned policy measures implemented by local government are mainly directed towards the population living in areas subject to urban renewal.

There needs to be a balance between physical, economic, and social regeneration of slum areas. Plans need to have a greater community involvement, not only in terms of consulting and informing, but also in terms of involvement in project definition and implementation. And this requires capacity building by community organisations. The opportunity to acquire professional qualifications, and the provision of adequate jobs, need to be central issues in these programmes. The involvement of private enterprises in local partnerships should also be a central objective, as should addressing the issue of equal opportunities in education, training and job access. As was noted earlier, to some extent local government is already doing this in the field of social policy.

Conclusions

One main conclusion is that Portugal, although mainly a country of emigration, has nevertheless a growing population of immigrants, albeit still very small in number, and in percentage terms, compared with most of the EU countries. The immigrant population is quite uniform regarding origin, being mainly from former Portuguese colonies, with Portuguese as the formal mother tongue, but with internal differentiation. The situation is more diverse if ethnic minorities in general are considered (including under this label the gypsy population).

Ethnicity is not yet a central question on the political agenda, and certainly not a central issue in local government policy, except in the LMA municipalities. Most of these have, already, special measures relating to ethnic minorities, and some have even created specific boards and councils to deal with the social integration process of these minorities. No special measures have been taken towards resident foreigners from more developed countries.

The increase of ethnicity's prominence in Portuguese society seems to be inevitable, and additional measures should be undertaken in order to improve the effectiveness of local government in Portugal. It is also necessary to strengthen local immigrants' associations (by providing premises, financial aid, professional training to directors and collaborators, etc.), in order to reinforce their capacity for intervention as well as their ability to enter into partnerships with municipalities, parishes, and NGOs. Yet in the absence of a clear and well structured national immigrant policy, intervention by individual authorities will not be sufficient.

Local government should concentrate on playing a supportive role for immigrants' associations, and in the creation of councils for the co-ordination of municipal and associations' actions. And these organisations should be more involved in the urban renewal policy framework from the beginning, in an approach we could classify as 'ethnic minority managerialism', in the sense that housing strategies and practices should put more emphasis on the distinctiveness of ethnic groups and cultures, and less on a universalistic concept of housing need.

Finally, it is felt that, in spite of national and local government and NGO efforts, it is still absolutely necessary to research deeply in this field. There are several aspects that should be researched concerning the everyday lives of immigrants and ethnic minorities. For example, what will be the long term consequences for the immigrant population, and for ethnic minorities in general, of a shift from a provider (local) state to a collaborative or partner (local) state? What will be the long term consequences, for ethnic minorities, of a shift from a policy of geographical co-operation to one of urban competition, where ethnic minorities will be a burden? Where should they be housed? How could such an area be marketed? What are the consequences of institutional racism, in the workplace and school, and in the process of local political representation? How should local government deal with ghettoes, and with emerging urban violence? These are some examples of questions for a future research agenda in Portugal.

Notes

1 By 'ethnicity' we mean the situation in which social inequality, cultural identity, and collective action, is a direct result of belonging to an ethnic minority group.

2 For a general characterisation of ethnic minorities in Portugal, see also Sant-Maurice, A., et al. (1989; 1995); and for communities of Indian origin, Malheiro, J. (1996).

3 It is difficult to separate the African population according to their legal situation: nationals, foreigners, and illegal immigrants. Recent research (CEPAC, 1995), in the districts of Lisbon and Setúbal, considered all residents with an African origin: all people with parents born in Africa, till the third generation. Cultural criteria were considered more important than legal status.

4 To establish the social contrast, the author considered residential location, age and sexual structure, scholarship, and social class composition; and for the cultural contrast, religion, language, race, marriage, and way of life.

5 One example is the recent legislation that allows immigrants' associations and other organisations to become prosecutor partners in cases of racist and xenophobia crimes (Law nº 20/96, 6-7-96).

6 Santos (1985) argues that there has never been a welfare state in Portugal, as there has in other parts of Europe.

Table 6.1 Number of immigrants

	Total immigrants	Africans	Africans from former colonies	Africans from colonies and Brazil
Total number	172,912	81,176	77,114	97,196
%	100	46.9		

Source: SEF, 1997.

Table 6.2 Number of immigrants from former Portuguese colonies (PALOP)

	Immigrants from PALOP	Cape Verde	Angola	Guinea-Bissau	Mozambique	São Tomé
Total number	77,114	39,546	16,282	12,639	4,413	4,234
%	100	51.3	21.0	16.4	5.7	5.5

Source: SEF, 1997.

References

Almeida, J.F. (1992), *Exclusão social. Factores e tipos de pobreza em Portugal* (*Social exclusion. Factors and types of poverty in Portugal*), Celta Editora, Lisbon.

Capucha, L. (1990), 'Associativismo e modos de vida num bairro de habitação social' ('Associations and way of life in social housing estates'), *Sociologia*, no. 8.

CE (1995), *Europa 2000+. Cooperação para o ordenamento do território (Europe 2000. Co-operation for urban planning)*, CE, Bruxelas.

CGTP (1993), *2º Encontro de Migrações (Second Meeting on Migration).*

C.M. Alcochete, *Relatório de Actividades (Activity Report, several years).*

C.M. Almada, *Plano de Actividades, 1997 (Activity Plan).*

C.M. Almada, *Relatório de Actividades (Activity Report, several years).*

C.M. Amadora, *Relatório de Actividades (Activity Report, several years).*

C.M. Azambuja, *Relatório de Actividades (Activity Report, several years).*

C.M. Barreiro, *Relatório de Actividades (Activity Report, several years).*

C.M. Cascais, *Relatório de Actividades (Activity Report, several years.*

C.M. Loures, *Relatório de Actividades (Activity Report, several years).*

C.M. Mafra, *Relatório de Actividades (Activity Report, several years).*

C.M. Moita, *Relatório de Actividades (Activity Report, several years).*

C.M. Palmela, *Relatório de Actividades (Activity Report, several years).*

C.M. Seixal, *Relatório de Actividades (Activity Report, several years).*

C.M. Sintra, *Relatório de Actividades (Activity Report, several years).*

C.M. V.F. Xira, *Relatório de Actividades (Activity Report, several years).*

Cunha, I.F. (1996), *Os africanos na imprensa portuguesa: 1993-1995 (The Africans in the Portuguese Press: 1993-1995)*, CIDAC, Lisbon.

DEDIAP/CEPAC (1995), 'Os números da imigração africana' ('The numbers of African immigration'), *Cadernos CEPAC*, no. 2, Lisbon.

DGAA (1995), *Administração Local em Números (Local Government in numbers).*

DGDR (1996), *Programas Urban e Reabilitação Urbana. Revitalização de áreas urbanas em crise (Urban and Urban Regeneration Programmes)*, DGDR, Lisbon.

Diário da Assembleia da República (Parliament Official Journal).

Diário da República (Official Journal).

Esteves, M. (1991), *Portugal, País de Imigração (Portugal, Country of immigration)*, IED, Lisbon.

França, I. et al. (1992), *A comunidade cabo-verdiana em Portugal (The Cape Verde community in Portugal)*, IED, Lisbon.

Haghighat, C. (1994), *L'Amérique urbaine et l'exclusion sociale*, PUF, Paris.

Harrison, M. and Law, I. (1997), 'Needs and empowerment in minority ethnic housing: some issues of definition and local strategy', *Policy and Politics*, vol. 25, no. 3, pp. 285-298.

Imrie, R. and Thomas, H. (1993), *British urban policy and the urban development corporations*, P. Chapman, London.

Imrie, R. and Thomas, H. (1995), 'Urban policy processes and the politics of urban regeneration', *International Journal of Urban and Regional Research (IJURR)*, vol. 19, no. 4, pp. 479-494.

INE (several years), *Estatísticas Demográficas (Demographic Statistics).*

INE, *Recenseamento da População, 1960 (Population Census, 1960).*

INE, *Recenseamento da População, 1981 (Population Census, 1981).*

INE, *Recenseamento da População, 1991 (Population Census, 1991).*

Machado, F. (1992), 'Etnicidade em Portugal: contrastes e politização' ('Ethnicity in Portugal: contrasts and politization'), *Sociologia*, no. 12.

Malheiro, J. (1996), *Imigrantes na região de Lisboa (Immigrants in the Lisbon Region)*, Ed. Colibri, Lisbon.

McGregor, A. and McConnachie, M. (1995), 'Social exclusion, urban regeneration and economic reintegration', *Urban Studies*, vol. 32 (no. 10), pp. 1587-1600.

Quedas, M.J. (1994), 'Expectativas habitacionais e coexistência de grupos étnicos' ('Housing expectation and coexistence of ethnic groups') *Sociedade e Território*, no. 20.

Rodrigues, W. (1989), 'Comunidade cabo-verdiana: marginalização e identidade' ('Cape-Vert community: marginalisation and identity'), *Sociedade e Território*, no. 8.

Sant-Maurice, A. et al. (1989), 'Descolonização e Migrações: os imigrantes dos PALOP em Portugal' ('De-colonisation and migration: the PALOP's immigrants in Portugal'), *Revista Internacional de Estudos Africanos*, no. 10-11.

Sant-Maurice, A. et al. (1995), 'Modes de vie des immigrants cap-verdiens résidant au Portugal', *Espaces et Sociétés*, no. 79.

Santos, B.S. (1985), 'Estado e sociedade na semiperiferia do sistema mundial: o caso português', ('State and society in the semi-periphery of the world system: the Portuguese case'), *Análise Social*, no. 87-88-89, pp. 869-901.

SEIES (1995), Associações de Imigrantes na Área Metropolitana de Lisboa (Immigrants' associations in the Lisbon Metropolitan Area), Lisbon.

Silva, C.N. (1993a), 'As determinantes económicas e políticas do planeamento municipal em Portugal' ('The economical and political determinants of municipal planning in Portugal') *Finisterra*, no. 55-56, pp. 23-58.

Silva, C.N. (1993b), 'Réformes concernant la coopération inter-municipale et les limites: nouvelles tendences au Portugal', *Bulletin de la Societe Languedocienne de Geographie, 1993*, no. 3-4, pp. 281-299.

Silva, C.N. et al. (1994), 'Le Pouvoir Local', *Peuples Méditerranéens*, vol. 66, pp. 91-102.

Silva, C.N. (1996a), 'O financiamento dos municípios', in César Oliveira (Coord.) *História dos Municípios e do Poder Local. Dos finais da Idade Média à União Europeia* (*Local Government Finance*), Círculo de Leitores, Lisbon, pp. 433-462 (ISBN 972-42-1300-5).

Silva, C.N. (1996b), 'Autarquias locais e Gestão do Território. Que diferença faz o partido político?' ('Local government and urban management: does politics matter?') *Finisterra*, no. 59-60, pp. 99-120.

Simões, C. et al. (org.) (1992), *Documentos do encontro 'A Comunidade Africana em Portugal' (The African community in Portugal - Documents)* Ed. Colibri, Lisbon.

Smith, D. et al. (1996), 'Citizenship, nationality and ethnic minorities in three European Nations', *International Journal of Urban and Regional Research (IJURR)*, vol. 20, no. 1.

SOS Racismo (1996*), Dossier de informação (Information Dossier).*

SOS Racismo (1996), *Proposta de lei contra a discriminação racial (Project of law against racial discrimination).*

SOS Racismo (n.d.), *Para uma política de integração social (For a policy of social integration).*

Tinhorão, J.R. (1988), *Os negros em Portugal - uma presença silenciosa (Blacks in Portugal - a silent presence),* Ed. Caminho, Lisbon.

Trindade, M.B.R. (1990), 'Migrações no Mercado Único Europeu' ('Migrations in the Single European Market'), *Análise Social*, no. 107, Lisbon.

Acknowledgements

This research is part of Project nº SSPS/C/PCL/2608/96 - '*Poder Local e Políticas Sociais*' *(Local Government and social policies)*, developed in CEG - Centro de Estudos Geográficos - University of Lisbon under the co-ordination of the author, and funded by JNICT-MSSS.

7 Urban renewal, social exclusion and ethnic minorities in Britain

HUW THOMAS

Introduction

This chapter considers the available evidence on the impact of urban renewal policies on black and ethnic minorities in the UK, setting this within an account of the country's governance and demography, and the evolution of urban policy. It argues that, although the socially constructed 'problem of race' has been close to the heart of the development of urban policy since the 1960s, political sensitivities about how (and to what extent) racism should be acknowledged has meant that policy has rarely (if ever) been targeted explicitly at the eradication of racism, and the collection of monitoring data has very rarely been relevant for assessing the impact of policy on black and ethnic minorities. Of necessity, therefore, there must be considerable reliance on a few case studies, sporadic evidence, and informed speculation. These admittedly imperfect sources provide no basis for optimism.

Institutions of government

Sub-national government in Britain has been the subject of two major reviews and reorganisations since the early 1970s (Wilson et al., 1994). These have led to changes in boundaries of local government units, and also to changes in the number of tiers of local government, and the division of responsibilities between tiers. The current position will be described shortly; but it is important to emphasise that there have been significant continuities within local government in the post-war period. First, and perhaps most significantly, local government functions and powers are defined in national legislation, and in the judicial interpretation of such legislation. Britain is a unitary (as opposed to federal) state, and it is the national parliament which confers powers on local government, and it is parliament which decides whether a review of functions, or reorganisation, is timely. Local government has no independent constitutional status, but - and this is the second important continuity - local government is, and has been, a major presence in urban and rural life, particularly since 1945. The significance of its presence can be measured in a number of ways. For example, we can point to the services for which various tiers of local government have been responsible at some time since 1945. The list would include education, social services, town planning, building and managing housing, drainage and refuse collection, fire services, water, and police. We might also note the sheer numbers of employees in local government. Wilson et al. (1994, pp. 228-229) claim that in 1990, 11 per cent of all full-time and part-time employment in the UK was in local government, despite central government attempts to restrict its size (indeed, employment growth in local government as a whole seems to have remained static from about 1980). A final important continuity is that most governmental functions at a local level have been undertaken by elected bodies: i.e. although organisations run by appointees of central government, or bodies shielded in some other way from direct/local accountability, are increasingly important (Cochrane, 1993), elected local government remains the most important presence in local delivery of public services (Stoker, 1991).

In speaking of continuities, however, the impression should not be given that British local government has remained essentially unchanged. Indeed, the continuities themselves have thrown up tensions. Most notably, the significant presence of elected local government - as a major employer dispensing large budgets on a wide range of services - has made it more or less impossible for local politicians and salaried council officers to view local councils as simply or solely delivery mechanisms for central government policies. Whatever the constitutional position of local government, historically its growth as a locally elected tier of government has been fuelled by local pressures (initially from middle class 'city fathers', later from more popular sources) for opportunities to put a local stamp on the way their town or city was run (Briggs, 1968). Locally elected councils can claim with some plausibility to 'speak for' their areas, and from time to time this has created major policy conflicts between central and local government, perhaps the most famous in the 1980s being the clashes between the left-wing Greater London Council (GLC) and the Conservative central government under Mrs Thatcher. The fact that the GLC, and other councils covering the major British conurbations, were abolished by central government in 1986, demonstrates how the ultimate governmental sanction can, on occasion, operate; and also illustrates the hostility and suspicion of local government which developed under successive Conservative governments after 1979.

Abolition of certain kinds of local councils was the most extreme expression of Conservative dislike of the prominence and (often) political complexion of local government. Throughout the 1980s and into the 1990s there has also been a whittling away of the powers and responsibilities of elected local government. Two main approaches have been taken in trying to curb local government: in some policy areas, councils have been obliged to allow outside agencies (usually private companies) to tender for various activities (for example, managing leisure centres); in other policy areas, central government has simply channelled its resources into agencies other than local government, and these agencies have been empowered to take on tasks previously undertaken by local councils. Thus, social housing for rent is now largely developed and managed by housing associations, rather than by departments of the local council. In urban renewal and urban regeneration, both of these strategies have been employed by central government. New agencies have been created (such as urban development corporations), while local councils have also been required to work with other organisations in the private and voluntary sectors (e.g. in City Challenge schemes) (Atkinson and Moon, 1994).

Central government initiatives and pressure have undoubtedly wrought changes in the roles and importance of local government. Cochrane (1993) asked 'Whatever Happened to Local Government?' and the phrase 'local governance' has come to denote an emerging system in which elected local government is less central than previously (Thomas and Imrie, 1997). There have certainly been changes, although, as previously stated, elected local councils remain the single most important agency of governance in their areas. Moreover, pressures for change in the nature of local government have by no means come only from central government. The dramatic local effects of economic restructuring, increasing professional awareness of the scope for innovation in management and organisation of public and private sector organisations, and the high political profile of 'consumer-power', have induced local councils to be innovative in their working practices (Stoker, 1991).

At present, therefore, local governance in Britain is institutionally the most complex it has been since 1945. Elected local councils remain central players, responsible still for large budgets, but increasingly choosing (and being obliged) to work with, and through, other organisations (public and private), to achieve their objectives. 'Enabling', rather than 'executing', has become the watchword of local government. The extent of the change can be exaggerated, but the direction of change is unmistakable.

These developments have great significance for Britain's ethnic minorities. The promotion of equal opportunities in recruitment and in service delivery has made steady (if unspectacular) progress in local government policy-making. New Right central government hostility to the principle and practice of promoting equal opportunities through specific policies had a dampening effect (Ball and Solomos, 1990), but a survey of about a hundred local authorities in the early 1990s still found that almost 70 per cent had explicit equal opportunities policies relating to recruitment (Krishnarayan and Thomas, 1993). The position in the new kinds of agencies increasingly involved in urban policy appears to be very different. Brownill et al. (1997), for example, found that very few urban development corporations (UDCs) had formal equal opportunities policies. Of course, having a formal policy of promoting equal opportunities does not guarantee that equal opportunities will be provided, but it is reasonable to suggest that formal policies mark a step forward in sensitising public policy to the reality of indirect and direct discrimination in economic and social life, and (potentially) in the delivery of public policy.

One reason why elected local authorities have gradually become more positive about sensitising policies and practices to the needs and aspirations of ethnic minorities may be that increasing numbers (and proportions) of elected councillors are drawn from ethnic minorities. The vast majority of ethnic minority councillors represent the Labour Party, but there are ethnic minority councillors in each of the three major national parties, and each party has made attempts to woo the 'ethnic vote' (Geddes, 1993; Saggar, 1993), and by doing so, has minimised the likelihood of ethnically based regional or national parties. Table 7.1 shows the latest available information in respect of proportions of councillors in England who are from ethnic minorities. The table makes it clear that the black and ethnic minority population is under-represented among local councillors; that ethnic minority women (especially Asian women) are particularly under-represented; and that there are considerably more Asian councillors than Afro-Caribbean councillors.

These differences are simply one illustration of a phenomenon which is attracting increasing scholarly and political attention, and which will be discussed in the section which follows: that different ethnic and racialised groups in Britain are following different socio-economic trajectories, making generalisation increasingly difficult, and potentially misleading.

Of particular relevance to this chapter is the finding of successive national surveys that certain minority ethnic groups are at least as well-represented among professional occupations as is the white population (Modood et al., 1997). Tables 7.2 and 7.3 provide data from a 1994 survey which illustrates this point. It appears, however, that public administration, which includes local government and central government, is not a major source of professional jobs, at least for ethnic minority men (see Modood et al., 1997). In this respect, the proportions of ethnic minorities in the planning profession is not exceptional, although the low proportions provide no grounds for complacency. A survey of the membership of the Royal Town Planning Institute conducted in 1988 found that under 3 per cent were from ethnic minorities (Nadin and Jones, 1990).

The black and ethnic minority population in the UK

There have been periodic influxes of migrants to Britain for many centuries, and although most of these have been from other European countries (notably Ireland), it is also the case that the country has long-established non-white minority ethnic populations. The pace of immigration, particularly of non-white people, picked up in the year following the Second World War, not least because of British government policies encouraging immigration to fill public sector jobs (Solomos, 1993). However, from the immediate post-war period onward, there has been a consistent racialisation of immigration in British politics, in the sense that non-white immigration has been portrayed as problematic (Solomos, 1993). From the 1950s there have been influential voices calling for control of non-white immigration, and from the 1960s, legal controls have indeed become increasingly tight, with a consequent fall in the numbers of immigrants.

The 1991 Census was the first to ask questions about ethnicity, and its findings will be used as a basis for discussing the ethnic minority population in the mid 1990s. There was a considerable discussion (and testing) of an appropriate question (or set of questions) in preceding years (Bulmer, 1996), and widespread acknowledgement that the question eventually used was sociologically imprecise, and represented a political compromise (using the term in the broadest sense) (Peach, 1996b). As Peach (1996b, pp. 5-6) commented, in a passage worth quoting at length:

> While birthplace is an unambiguous category, ethnic identity is more mercurial. Critically, ethnicity, is contextual rather than absolute. One may be Welsh in England, British in Germany, European in Thailand, White in Africa. A person may be Afro-Caribbean by descent but British by upbringing so that his or her census category might be either Black-Caribbean or Black-Other. Similarly, a person may be an East African Asian, an Indian, a Sikh or Ramgarhia. Thus ethnicity is a situational rather than an independent category. The Census Quality Validation has indicated confusion about which category to claim, especially for those of mixed ethnic background.
>
> The ten main ethnic categories produced by the 1991 census are not unambiguous. The emphasis in the classification is on categorising the non-European minority groups, while aggregating the European population into a single White group. Taking the categories in the order in which they appear in the census, White is not an ethnic group but a racial designation; it conflates all European identities and more besides. Black-Caribbean is complicated by the uncertainty as to whether those who wrote themselves into the census as 'Black British' ... should have been included in this category rather than Black-Other. Within the Black-Caribbean group there are differences between those originating from different islands and territories within the Caribbean. Black-African is clearly a racial designation, but it hides a great range of national and ethnic identities within it and does not represent a homogeneous group. Black-Other is similarly a residual category, containing not only second generation Afro-Caribbeans, but also African-Americans (significant in the geography of military bases) and old established dockland ethnic minorities in Liverpool, Cardiff and other places ... Indian as a category represents nationality as much as ethnicity. There are very significant differences within the group according to religion, language and place of origin ... Pakistani is relatively unambiguous as a title, although significant differences exist within the group according to place of origin ... Bangladeshi is perhaps the least ambiguous of the groups, with a high proportion originating from the Sylhet District in the north east of the country ... The Chinese are drawn from Hong

Kong, Singapore, Vietnam and Taiwan as well as from the People's Republic itself ... Other-Asians are another residual category containing the Japanese, Malaysians and a myriad of small groups ... Other-Other is perhaps the most ambiguous of all the groups, although Middle Easterners figure prominently within it. Finally, the Irish pose a significant problem of identification. There was no Irish ethnic question on the census and birthplace has additional problems of interpretation in this case as two very different traditions are involved for the 'born in Ireland' category ....

These comments would be enough in themselves to alert us to the complexity of interpreting census results. But in relation to ethnicity there is an additional problem caused by under-enumeration (Simpson, 1996). In brief, the overall response rate to the census was 97.8 per cent, but the level of enumeration 'was significantly lower among young children, young adults, and very elderly people, in particular among men aged 20-34 in city districts' (Simpson, 1996, p. 63). These deficiencies were more significant for some ethnic groups than others. On the basis of comparison of 1991 census returns with other data sources (including analyses of earlier censuses), Simpson (1996) argues that it is likely that there is a 'serious underestimate' (p. 76) of thc numbers of ethnic minority group children, and that there was also disproportionate under-enumeration of ethnic minority adults aged 15-34 (though the size of this effect is unquantifiable).

The 1991 census must be interpreted cautiously, therefore, even as a snapshot of the ethnic composition of Britain. In this section, its findings will be used simply to provide a very broad picture of what calling Britain a multi-ethnic or multi-racial country might mean.

Table 7.4 shows the ethnic breakdown of the population, in broad terms, in 1991, with Table 7.5 providing an estimate of the growth of the larger minority ethnic groups in the post-war period. Owen (1996, p. 82) estimates - on the basis of incomplete data - that the ethnic minority population of Great Britain has increased from about 886,000 in 1966; as Peach (1996b) points out, although minority ethnic groups remain a small proportion of the total population, their rate of growth has been relatively fast.

Two other broad generalisations can be made on the basis of census data, and both are supported by other survey data. Firstly, that in certain respects the white population is quite different from the ethnic minority population, more or less taken as a whole; but secondly, that the ethnic minority population is by no means homogenous, and there appear to be significant variations *between* minority ethnic groups in relation to some socio-economic characteristics.

Perhaps the two most significant ways in which ethnic minority groups, in general, differ from the (white) ethnic majority are (i) the distribution of their populations by age, and (ii) their spatial distribution.

Age/sex pyramids for selected ethnic groups make apparent the relative youth of all the 'non white' ethnic groups (though their pyramids are by no means uniform) (See Peach, 1996b, pp. 12-13). A young population has very distinctive needs, for example, in relation to health and welfare, education, and recreation; and meeting these needs has land-use implications. So it is plausible to suggest that the ethnic minority population has a special interest in public policies (including planning policies) which have a bearing on health and welfare, education, and recreation.

Tables 7.6 and 7.7 illustrate the differences in the geographical distribution of the British population in 1991. The ethnic minority population is concentrated in England: and within England, in very few regions. In addition the ethnic minority population is concentrated in large urban areas, and is over-represented in what Owen (1995) has termed 'declining industrial centres'. This concentration is repeated at smaller spatial scales, where a degree of residential segregation by ethnicity is apparent. Peach (1996a; 1996b) points out that nowhere in Britain is there to be found residential segregation of the kind experienced 'almost uniformly' by African-Americans; however, a degree of ethnic segregation which cannot be explained as a function of class undoubtedly exists. An Index of Dissimilarity represents 'the percentage of the population which would have to shift from its area of residence in order to replicate the distribution of the total population in the city. It has a range from 0 (no segregation) to 100 (total segregation)' (Peach, 1996c, p. 37). On the basis of such an index, the Bangladeshi, Pakistani, Black-African, and Chinese populations (in that order), appear to be the most segregated (from all other groups) at the fine-grained enumeration district level, while there is some evidence that the Black-Caribbean population of Greater London is undergoing a 'progressive dispersal' (Peach and Rossiter, 1996, p. 124).

Using a different measure of segregation, a measure of the exposure of one group to another, Peach and Rossiter (1996, p. 126) provide a revealing insight into aspects of the social reality of residential segregation:

The high degree of Bangladeshi clustering in London means that they have a 24 per cent chance of meeting other Bangladeshis in enumeration districts where they live, although they have twice as great a chance of meeting whites. Whites in London have only a 1 per cent chance of living in an enumeration district which contains Bangladeshis.

It is important to appreciate the force of Peach's comparisons between residential segregation in the USA and UK; nevertheless, it remains much more likely for members of ethnic minorities than it is for white people that they will live in fairly close proximity to members of a minority group: either the same minority group as themselves, or another minority group. Though propinquity does not entail community, it is difficult to doubt that a lack of regular visual contact and neighbourly dealings can do anything but bolster the unease which many white people have in their dealings with members of the non-white population (Hall, 1978).

Turning now to some ways in which the ethnic minority population is itself heterogeneous, a recent authoritative analysis of census data, and a specially commissioned national survey, concluded that:

> On many measures of education, employment, income, housing and health, there is a two- or three-way split, with Chinese, African-Asian and sometimes Indian people in a similar position to whites, Caribbeans some way behind, and Pakistanis and Bangladeshis a long way behind them.
>
> (Modood et al., 1997, p.10.)

By way of illustration, Tables 7.8 and 7.9 show the relationship between ethnicity and socio-economic class, for men and women. A number of features are worthy of mention. Firstly, that certain ethnic minority groups have far and away the highest proportions in professional classes - see, for example, the Chinese, Black-African and Indian groups. Secondly, the proportions of economically active women, in particular, vary widely, with Pakistani and Bangladeshi women much less likely to be engaged in formal employment than women of other ethnic groups. Among the economically active, the proportions in employment differ quite widely between various ethnic groups. Again, for women and (especially) men, Pakistanis and Bangladeshis stand out as having lower rates of employment, while members of the white, Irish, Chinese, Indian, and (for women) Black-Caribbean groups stand out as having high employment rates. Chinese and South Asian groups have a much higher proportion of their members in self-employment, and two thirds of Bangladeshi male residents over the age of 16, and 60 per cent of Chinese males in the same age group, were either self employed or employees, in distribution or catering (Peach 1996b, p. 18). Variations such as these show the dangers of generalising about the impact of any public policy (including urban renewal policies) on ethnic minorities.

Nevertheless, Mason (1995, p. 51) has commented that 'A long series of studies of the employment status of members of minority ethnic groups has shown that in general terms they are employed in less skilled jobs, at lower job levels, and are concentrated in particular industrial sectors'. As we have seen, the male members of some minority ethnic groups are exceptions to this generalisation, but there is 'overwhelming evidence', nevertheless, 'that discrimination is a continuing and persistent feature of the experience of Britain's citizens of minority ethnic origin' (Mason, 1995, p. 58). Mason summarises the findings of a series of studies from the 1970s to today which have found significant levels of *direct* or deliberate discrimination. However, there is also evidence of widespread *indirect* discrimination, for example where selection criteria for recruitment are applied across the board, but put one or more ethnic group(s) at a systematic disadvantage. Racial discrimination alone cannot explain the complexity of the pattern outlined above, but it remains a significant factor in the lives of many ethnic minority residents of Britain.

Urban policy frames and post 1945 urban renewal

For the purposes of this chapter, 'urban renewal' will be understood in the broad sense set out by Gibson and Langstaff (1982, pp. 12-13): the redevelopment and rehabilitation of older parts of towns and cities, with low property and land values, in pursuit of a variety of policy objectives (including, most prominently, creating areas of new commercial development, environmental improvements, and areas of new housing, including publicly owned housing for rent). In this sense, urban renewal is one part of urban policy, which, following Atkinson and Moon (1994, ch. 1), we recognise to be a loose collection of government initiatives aimed at the alleviation of social, economic and physical ills concentrated in (though not necessarily exclusive to) urban areas. At times, however, the main (indeed almost exclusive) focus of British urban policy has been urban renewal, as physical redevelopment and regeneration has been viewed as the key to better towns and cities.

Thus, throughout the 1950s and into the 1960s, the wholesale clearance of areas of building (generally housing) deemed 'obsolescent' was undertaken for two main purposes: city centre redevelopment (Ambrose and Colenutt, 1975), and 'slum clearance' (ie replacement of sub-standard housing by new houses to rent, sometimes in the same area, but often supplemented by housing built on the periphery of towns and cities). For this reason it is a little misleading to suggest, as Atkinson and Moon (1994) do, for example, that urban policy was a product of the 1960s. However, what is the case - and this will be discussed shortly - is that in the 1960s, public policy became concerned about poverty and aspects of urban life.

City centre redevelopments were usually undertaken jointly by local authorities and private developers, but slum clearance by local authorities alone. In both cases, however, the local authority was responsible for land acquisition in order to create sites which were economic to develop, a process which almost inevitably involved an element of compulsory acquisition. Poor residents of run-down areas had little leverage over determined local authorities (Dumbleton, 1977). Once they lost their homes they found themselves at the mercy of local authority housing and letting policies which very often scattered residents from long-standing communities to a variety of housing estates. This outcome should not be seen as entirely unintentional. Wilfred Burns, one of the leading planners of the 1960s, commented:

> this is a good thing when we are dealing with people who have no initiative or civic pride. The task, surely, is to break up such groupings even though the people seem to be satisfied with their miserable environment and seem to enjoy an extrovert social life in their own locality.
>
> (Burns, 1963, pp. 94-95; quoted in Ward, 1994, p. 153.)

Here we have a conception of urban renewal as imposing a kind of social hygiene on cities, and a linking of physical and social objectives which is reminiscent of the pioneers of housing standards of a century earlier. The disruptive effect of clearance has been recognised by ethnic minority communities, many of whom have mobilised to oppose wholesale clearance in favour of grants for improvements to existing properties. However, the lack of political leverage endured by minority communities means that stories of successful opposition are rare (see, e.g., Stoker and Brindley, 1985), and those of frustration more common (Heywood and Naz, 1990).

As disruptive and damaging as actual clearance and rehousing, could be the blight caused by inclusion in a clearance or redevelopment programme, even though there might be no immediate prospect of compulsory acquisition. The blight could take many forms. One was the impossibility of receiving government grants, or loans from building societies or other reputable sources, for home improvements or for house purchase (making the properties very difficult to sell). Another common form was a high turnover of population, as those who could afford to leave (for example, by renting more expensive properties, or by accepting a loss on a house that was owned) did so, and a more transient population replaced them. By the late 1960s, physical redevelopment and housing clearance had a reduced role in urban policy, with general environmental upgrading (including, but going beyond, the refurbishment of older housing) becoming more significant. However, the focus of urban policy was itself broadening, with the so-called 'rediscovery of poverty' from the mid 1960s (see, for example, the classic text by Coates and Silburn, 1970). In 1968 a Labour government launched a programme which was to be a key component of urban policy for the more or less the next twenty years: the Urban Programme (UP) (originally entitled 'Urban Aid'). The programme was to have major reorientations during its existence (see below), but certain elements remained constant, and defined a political and professional consensus in respect of how urban policy should be conceived and delivered.

Firstly, the resources targeted under the UP were small in scale compared with mainstream funding of education, social services, health, etc., and were intended to meet 'special needs'. The targeting was spatial - a list of eligible local authority areas was drawn up - and there was a consistent tension between this spatial targeting and the realisation that the 'special needs' being addressed extended beyond the target areas. Nevertheless, the area-based approach remains central to British urban policy, even after the demise of the UP.

Secondly, elected local authorities had pivotal roles in the UP. UP funds were channelled through local authorities, and central government and UP funding had to be supplemented by local authority funding (in the ratio 3:1). In effect, then, eligible projects had to satisfy both central government and local government criteria. Most projects were actually sponsored and undertaken by local authorities themselves, though some were undertaken by voluntary and community organisations. The private sector was not involved.

Finally, 'the subliminal presence of race [was a] factor in structuring the development of programmes' (Atkinson and Moon, 1994, p. 234). Assisting ethnic minorities became an explicit objective of the UP from 1974, but long before that, the notion of 'the urban', and especially of the 'inner city', had been constructed in a racialised fashion (Brownill and Thomas, 1998; Solomos, 1993). To this day, urban policy is affected by the wider politics of race, and its focus is often muddied by tensions, between New Right hostility to positive action to assist ethnic minorities, and the long-standing ambivalence of centrist and leftist parties towards acknowledging the seriousness of racism. As a result, as Harrison (1989, p. 51) notes:

> Although race is on the agenda as an aspect of policy, it is ... very much subordinate to the goals of general economic and environmental regeneration.

These three elements structured urban policy throughout the 1970s. By the end of the decade they were supplemented by a further factor: the belief that the kinds of problems which urban policy was targeting were manifestations of structural changes in the economy. Hailed by some commentators as 'the first serious attempt by a government in the post-war era to understand the nature and causes of Britain's urban problems' (Atkinson and Moon, 1994, p. 66), the 1977 government White Paper 'Policy for the Inner Cities' summed up an emerging political and academic consensus that economic restructuring, international in scale, was the root cause of the de-industrialisation, and consequent unemployment, population decline, social malaise, and physical decay, of certain parts of British cities. The government's response involved intensified spatial targeting of urban policy, an attempt to broaden the 'partners' involved in local policy formulation and delivery (to include the private sector and voluntary groups, while retaining a key role for elected local authorities), and a shift from funding 'social' projects (play schemes for children, for example) to funding 'economic' schemes (workshops for new businesses might be a typical example). It is debatable whether the policy responses matched the diagnosis of the problem (Atkinson and Moon,

1994), and overall urban policy in the late 1970s and into the early 1980s had a similar 'feel' to it as it did ten years earlier.

This could hardly be said of urban policy in the late 1980s. The UP has been gradually diminishing in importance over a number of years, and is due to expire. But more important than the demise of a particular programme was the new policy frame within which urban problems were being defined and addressed (Solesbury, 1987). From the early 1980s onwards, successive Conservative governments promoted, with increasing insistence, a set of policies and initiatives which were informed by the following principles. Firstly, the primary objective of urban policy should be to facilitate market-based solutions to urban problems. Secondly, in practice, this meant facilitating property development (certainly for most of the 1980s): the so-called 'property-led' approach (Healey et al., 1992). Thirdly, scarce resources for urban policy implementation should be distributed by competition rather than by an index of need, as a way of encouraging better thought out projects. Finally (but by no means least significantly), the role of elected local authorities in formulating and delivering urban policy should be drastically diminished, and in particular, private sector perspectives should be given a greater prominence. Achieving this latter task has involved a number of devices, perhaps the primary one being the setting up of large numbers of single-purpose, short-term organisations to deliver particular strands of urban policy - urban development corporations, (UDCs), City Challenge teams, and Single Regeneration Budget (SRB) projects, to name but a few. Elected local authorities are involved in some of these to varying extents, but they never have the centrality they had with the UP, while private sector organisations are also involved to a far greater extent.

These twists and turns in urban policy have been exhaustively reviewed in the literature (e.g. Atkinson and Moon, 1994; Hambleton and Thomas, 1995) and in this chapter it will suffice simply to note some of their apparent implications for Britain's ethnic minority population. The first point to make is that assessing the implications is not easy, as very little evaluation of this dimension of urban policy has been undertaken, nor has data been collected with a view to monitoring the impact of policy on ethnic minorities (Harrison, 1989; Brownill et al., 1997; Russell et al., 1996). A major government-funded evaluation of urban policy has virtually nothing to say about its effects on black and ethnic minorities (Robson et al., 1994), an omission which is consistent with the ambivalence of urban policy in relation to acknowledging ethnic diversity, let alone widespread and persistent racism, in British society. What little evidence is available suggests that ethnic minorities were (and are) more likely to benefit from urban policy if certain conditions are met: firstly, that policy explicitly addresses combating racism and racial disadvantage (Ollerearnshaw, 1988; Munt, 1994); secondly, that funding leans towards 'social' projects, and revenue expenditure (i.e. salaries, running costs) (Munt, 1994), rather than 'economic' projects of a capital nature (buildings, etc) (Ball, 1988; Harrison, 1989); and, thirdly, that projects and local policy delivery are not in the hands of short-term, 'quick fix' agencies working to tight deadlines (Brownill et al., 1997). Few of these conditions have held for any length of time in any single place in Britain.

There is, however, a little indirect evidence that some ethnic minority groups were disproportionately affected by the uncertainty caused by the ambitious 1980s urban renewal projects described earlier. Gibson and Langstaff (1982), for example, note that in Birmingham, though the first wave of clearance areas tended not to have large proportions of ethnic minority residents, the adjoining areas, often designated for future action, did. As Ratcliffe (1992) comments, such designations typically made it difficult to raise loans on property and to sell them. Ratcliffe's own research relates to the 1980s, by which time urban renewal had come to mean grant-aided improvement of housing and the environment more generally, and many local authorities, such as Birmingham, had employed 'liaison officers' to facilitate communication between themselves and ethnic minority communities.

Ratcliffe's (1992) conclusion, however, is that the local authority had no serious intention of engaging ethnic minorities in decision-making over urban renewal, and the liaison officers were viewed as mere conduits for information. Urban renewal was still conceived by council officers as an essentially technical operation and the (generally non-technical) liaison officers were, therefore, marginalised in decision-making processes. Ratcliffe's findings report on the situation in the late 1980s, but subsequent work by Krishnarayan and Thomas (1993) in relation to the planning system in general, and Brownill et al. (1997) in relation to urban development corporations (UDCs), suggests that nothing of substance has changed. Though these research projects do not focus directly on the material impacts of urban renewal on ethnic minorities, their findings that ethnic minority voices are marginal in the renewal process do not give any reason for optimism about what those impacts might be.

Future research

It is, however, on the material impacts of urban renewal that future research needs to concentrate. As Moore (1992) has argued, spatial variations in the racialisation of social relations mean that generalisations and extrapolations about the dynamics of the interaction between urban renewal policies and the racialised labour and housing markets will always be insecure, if based on only a few case studies. A large number of case studies are needed, therefore; to date, there are virtually none.

Complementing detailed case studies, there should be thorough, competent, and standardised evaluations of all urban policy initiatives. This point goes well beyond the issue of impacts on ethnic minorities. Russell et al.'s (1996) evaluation of the City Challenge initiative is only the latest to point out that individual projects were collecting data for evaluation in an entirely non-standard way, making comparisons impossible. Base-line sets of indicators or variables can be identified only in the light of models of the workings of (and articulation of) local labour and housing markets, and the potential for interaction between these markets and urban renewal policies. In order for such models to be useful in relation to understanding the potential implications of policies on ethnic minorities, they would have to incorporate an explicit recognition of racism and discrimination in markets, and in social life more generally. As Munt (1994) has pointed out, policy makers have signally failed to do this to date, though some of the relevant research that would allow them a firm analytical basis for doing so has been undertaken.

Table 7.1 Asian and Afro-Caribbean local authority councillors by party in England 1992

	Total	Labour	Conservative	Liberal Democrats	Independent
Afro-Caribbean men	53	48	1	3	1
Afro-Caribbean women	32	28	2	0	2
Asian men	185	160	10	7	7
Asian women	17	12	2	2	0
	287	248	15	12	10

Source: Tables 3-6 of Geddes (1993).

Table 7.2 Job levels of men (base: male employees and self-employed)

(column percentages)

Socio-economic group	White	Caribbean	Indian	African Asian	Pakistani	Bangladeshi	Chinese
Prof./Managers/Employers	30	14	25	30	19	18	46
Employers and Managers (large establishments)	*11*	*5*	*5*	*3*	*3*	*0*	*6*
Employers and Managers (small establishments)	*11*	*4*	*11*	*14*	*12*	*16*	*23*
Professional workers	*8*	*6*	*9*	*14*	*4*	*2*	*17*
Intermediate and junior non-manual	18	19	20	24	13	19	17
Skilled manual and foreman	36	39	31	30	46	7	14
Semi-skilled manual	11	22	16	12	18	53	12
Unskilled manual	3	6	5	2	3	3	5
Armed forces or N/A	2	0	3	2	2	0	5
Non-manual	**48**	**33**	**45**	**54**	**32**	**37**	**63**
Manual	**50**	**67**	**52**	**44**	**67**	**63**	**31**
Weighted count	789	365	349	296	182	61	127
Unweighted Count	713	258	356	264	258	112	71

Source: Modood et al. (1997), p. 100.

Table 7.3 Job levels of women in work (base: female employees and self-employed)

(column percentages)

	White	Caribbean	Indian	African Asian	Pakistani	Chinese
Professional, managerial and employers	16 (15)*	5 (5)	11 (7)	12 (10)	12 (6)	30 (25)
Intermediate non-manual	21	28	14	14	29	23
Junior non-manual	33	36	33	49	23	23
Skilled manual and foreman	7 (2)	4 (2)	11 (3)	7 (3)	9 (3)	13 (-)
Semi-skilled manual	18	20	27	16	22	9
Unskilled manual	4	6	4	1	4	2
Armed forces/ inadequately described/ not stated	0	1	1	1	0	0
Non-manual	**70**	**69**	**58**	**75**	**64**	**76**
Manual	**29**	**30**	**42**	**24**	**35**	**24**
Weighted count	734	452	275	196	60	120
Unweighted count	696	336	260	164	64	63

* The figures in parentheses are exclusive of self-employed.

Source: Modood et al. (1997), p. 104.

Table 7.4 Population of Great Britain by Census ethnic groups, 1991

Ethnic group	Population
White	51,874,000
Black Caribbean	500,000
Black African	212,000
Black Other	178,000
Indian	840,000
Pakistani	477,000
Bangladeshi	163,000
Chinese	157,000
Other Asian	198,000
Other groups	290,000

Source: 1991 census, as reported in Modood et al. (1997), p. 13.

Table 7.5 Estimated size and growth of the Caribbean, Indian, Pakistani and Bangladeshi ethnic populations in Great Britain, 1951-91

	West Indian or Caribbean	Indian	Pakistani	Bangladeshi
1951	28,000	31,000	10,000	2,000
1961	210,000	81,000	25,000	6,000
1966	402,000	223,000	64,000	11,000
1971	548,000	375,000	119,000	22,000
1981	545,000	676,000	296,000	65,000
1991	500,000	840,000	477,000	163,000

Source: Peach (1996b), p. 9.

Table 7.6 Ethnic population by standard regions, Great Britain 1991

Region	Total	Per Cent of Great Britain	Minority	Per Cent of Minority
North	3,026,732	5.5	38,547	1.3
Yorks and Humberside	4,836,524	8.8	214,021	7.1
East Midlands	3,953,372	7.2	187,983	6.2
East Anglia	2,027,004	3.7	43,395	1.4
South East	17,208,264	31.3	1,695,362	56.2
South West	4,609,424	8.4	62,576	2.1
West Midlands	5,150,187	9.4	424,363	14.1
North West	6,243,697	11.4	244,618	8.1
Wales	2,835,073	5.2	41,551	1.4
Scotland	4,998,567	9.1	62,634	2.1
Great Britain	54,888,844	100.0	3,015,050	100.0

Source: Peach (1996b), p. 11.

Table 7.7 Relative concentration of ethnic minority population in selected metropolitan counties, Great Britain, 1991

	Total	White	Black-Caribbean	Black-African	Black-Other	Indian	Pakistani	Bangladeshi	Chinese
Great Britain	54,888,844	51,873,794	499,964	212,401	178,401	840,255	476,555	162,835	156,938
Greater London	6,679,699	5,333,580	290,968	163,635	80,613	347,091	87,816	85,738	56,579
West Midlands metropolitan county	2,551,671	2,178,149	72,183	4,116	15,716	141,359	88,268	18,074	6,107
Greater Manchester metropolitan county	2,499,441	2,351,239	17,095	5,240	9,202	29,741	49,370	11,445	8,323
West Yorkshire metropolitan county	2,013,693	1,849,562	14,795	2,554	6,552	34,837	80.540	5,978	3,852
Percentage ethnic group in named areas	25.04	22.58	79.01	82.66	62.83	65.82	64.21	74.45	47.70

Source: Peach (1996b), p. 11.

Table 7.8 Ethnic group for men aged 16 or over, by socio-economic class Great Britain, 1991

	Economically active	In Employment	I	II	III Non-Manual	III Manual	IV	V
Total	73.3	87.4	6.8	27.4	11.5	32.2	16.4	5.7
White	73.2	88.0	6.7	27.6	11.3	32.4	16.3	5.7
Black-Caribbean	80.1	73.8	2.4	14.2	12.2	38.9	23.6	8.7
Black-African	69.0	66.8	14.3	24.5	17.5	17.6	17.3	8.9
Black-Other	81.9	70.5	3.2	24.8	17.2	30.2	17.6	7.1
Indian	78.1	84.9	11.4	27.2	14.4	23.8	18.1	4.0
Pakistani	73.3	68.9	5.9	20.3	13.5	29.9	24.1	6.3
Bangladeshi	72.4	67.3	5.2	8.5	12.9	31.5	35.0	6.8
Chinese	70.1	88.1	17.6	23.3	19.3	29.5	8.0	2.4
Other-Asian	76.2	83.0	15.9	34.3	18.2	16.0	12.3	3.3
Other-Other	75.4	77.7	14.5	30.5	16.1	19.8	14.0	5.1
Irish-born	70.1	84.3	6.9	23.3	7.7	34.3	16.9	10.9

Source: Peach (1996b), p. 16.

Table 7.9 Ethnic group for women aged 16 and over, by socio-economic class Great Britain, 1991

	Economically active	In Employment	I	II	III Non-Manual	III Manual	IV	V
Total	49.9	92.1	1.7	25.9	38.8	7.6	18.0	7.9
White	49.7	92.6	1.7	25.9	39.0	7.6	17.8	8.0
Black-Caribbean	66.9	84.1	1.0	30.3	33.7	6.9	19.5	8.5
Black-African	60.1	71.0	3.0	31.8	30.8	5.6	16.9	12.0
Black-Other	62.9	77.9	1.3	25.2	40.6	9.3	19.1	4.6
Indian	55.4	85.3	4.4	20.9	34.9	6.4	29.2	4.1
Pakistani	27.1	65.5	2.7	22.3	34.2	6.5	31.7	2.6
Bangladeshi	21.8	57.6	1.8	22.9	35.8	6.4	26.6	6.4
Chinese	53.1	90.0	7.6	28.5	31.6	13.0	13.9	5.4
Other-Asian	53.9	84.9	6.0	30.7	33.8	7.0	16.6	6.0
Other-Other	53.9	82.5	4.5	30.8	38.4	6.2	15.1	4.9
Irish-born	49.9	92.4	2.6	33.2	26.8	6.3	18.5	12.6

Source: Peach (1996b), p. 18.

References

Ambrose, P. and Colenutt, B. (1975), *The Property Machine*, Penguin, Harmondsworth.

Atkinson, R. and Moon, G. (1994), *Urban Policy in Britain*, Macmillan, London.

Ball, H. (1988), 'The limits of influence: ethnic minorities and the partnership programme', *New Community*, vol. 15, no. 1, pp. 7-22.

Ball, W. and Solomos, J. (ed.) (1990), *Race and Local Politics*, Macmillan, London.

Briggs, A. (1968), *Victorian Cities,* Penguin, Harmondsworth.

Brownill, S. et al. (1997), *Race Equality and Local Governance*, ESRC Project Paper No. 3, Department of City and Regional Planning, UWCC, Cardiff.

Brownill, S. and Thomas, H. (1998), 'Urban Policy Deracialised?', in Yiftachel, O. et al. (eds) *The Power of Planning*, Kluwer, Netherlands.

Bulmer, M. (1996), 'The ethnic group question in the 1991 Census of population', in Coleman, D and Salt, J (eds) *Ethnicity in the 1991 Census*, Volume One, HMSO, London.

Burns, W. (1963), *New Towns for Old*, Leonard Hill, London.

Coates, K. and Silburn, R. (1970), *Poverty: The forgotten Englishmen*, Penguin, Harmondsworth.

Cochrane, A. (1993), *Whatever Happened to Local Government?*, Open University Press, Buckingham.

Dumbleton, B. (1977), *The Second Blitz*, Dumbleton, Cardiff.

Geddes, A. (1993), 'Asian and Afro-Caribbean representation in elected local government in England and Wales', *New Community*, vol. 20, no. 1, pp. 43-57.

Gibson, M. and Langstaff, M. (1982), *An Introduction to Urban Renewal*, Hutchinson, London.

Hall, S. (1978), 'Racism and Reaction', in Commission for Racial Equality *Five Views of Multi-Racial Britain*, CRE, London.

Hambleton, R. and Thomas, H. (eds) (1995), *Urban Policy Evaluation*, Paul Chapman Publishing, London.

Harrison, M.L. (1989), 'The Urban Programme, Monitoring and Ethnic Minorities', *Local Government Studies*, vol. 15, no. 4, pp. 49-64.

Healey, P. et al. (eds) (1992), *Rebuilding the City*, E & F.N. Spon, London.

Heywood, F. and Naz, M. (1990), *Clearance: the view from the Street*, Community Forum, Birmingham.

Krishnarayan, V. and Thomas, H. (1993), *Ethnic Minorities and the Planning System*, RTPI, London.

Mason, D. (1995), *Race and Ethnicity in Modern Britain*, OUP, Oxford.

Modood, T. et al. (1997), *Ethnic Minorities in Britain*, PSI, London.

Moore, R. (1992), 'Labour and housing markets in inner city regeneration', *New Community*, vol. 18, no. 3, pp. 371-386.

Munt, I. (1994), 'Race, urban policy and urban problems: a critique on current UK practice', in Thomas, H. and Krishnarayan, V. (eds), *Race Equality and Planning: Policies and Procedures*, Avebury, Aldershot.

Nadin, V. and Jones, S. (1990), 'A Profile of the Profession', *The Planner*, vol. 76, no. 3, pp. 14-24.

Ollerearnshaw, S. (1988), 'Action on equal opportunities in inner cities: the need for a policy commitment', *New Community*, vol. 15, no. 1, pp. 31-46.

Owen, D. (1995), 'The spatial and socio-economic patterns of minority ethnic groups in Great Britain', *Scottish Geographical Magazine*, vol. 11, no. 1, pp. 27-35.
Owen, D. (1996), 'Size, structure and growth of the ethnic minority populations', in Coleman, D and Salt, J. (eds) *Ethnicity in the 1991 Census*, Volume One, HMSO, London.
Peach, C. (1996a), 'Does Britain have ghettoes?', *Transactions of the Institute of British Geographers*, vol. 21, no. 1, pp. 216-235.
Peach, C. (1996b), 'Introduction', in Peach, C (ed.) *Ethnicity in the 1991 Census*, Volume Two, HMSO, London.
Peach, C. (1996c), 'Black-Caribbeans: class, gender and geography', in Peach, C. (ed.), *Ethnicity in the 1991 Census*, Volume Two, HMSO, London.
Peach, C. and Rossiter, D. (1996), 'Level and nature of spatial concentration and segregation of minority ethnic populations in Great Britain, 1991' in Ratcliffe, P. (ed.), *Ethnicity in the 1991 Census*, Volume Three, HMSO, London.
Ratcliffe, P. (1984), *Racism and Reaction*, Routledge and Kegan Paul, London.
Ratcliffe, P. (1992), 'Renewal, regeneration and 'race': issues in urban policy', *New Community*, vol. 18, no. 3, pp. 387-400.
Robson, B.T. et al. (1994), *Assessing the Impact of Urban Policy*, HMSO, London.
Russell, H. et al. (1996), *City Challenge. Interim National Evaluation*, The Stationery Office, London.
Saggar, S. (1993), 'Competing for the black vote', *Politics Review*, vol. 2, no. 4, pp. 26-31.
Simpson, S. (1996), 'Non-response to the 1991 Census: the effect on the Ethnic group enumeration', in Coleman, D and Salt, J (eds) *Ethnicity in the 1991 Census*, Volume One, HMSO, London.
Solesbury, W. (1987), 'Urban policy in the 1980s: the issues and arguments', *The Planner*, vol. 73, no. 6, pp. 18-22.
Solomos, J. (1993), *Race and Racism in Britain*, 2nd ed. Macmillan, London.
Stoker, G. (1991), *The Politics of Local Government*, 2nd ed., Macmillan, London.
Stoker, G. and Brindley, T. (1985), 'Asian Politics and Housing Renewal', *Policy and Politics*, vol. 13, no. 3, pp. 281-303.
Thomas, H. and Imrie, R. (1997), 'Urban Development Corporations and Local Governance in the UK', *Tijdschrift voor economische en sociale geografie*, vol. 88, no. 1, pp. 53-64.
Ward, S.V. (1994), *Planning and Urban Change*, PCP, London.
Wilson, D. et al. (1994), *Local Government in the United Kingdom*, Macmillan, London.

Acknowledgements

Some of the research on which this chapter is based was undertaken with Sue Brownill, of Oxford Brookes University, in a project funded by the (UK) Economic and Social Research Council, grant no. L311253060.

8 Comparative perspectives and research agenda

ABDUL KHAKEE, PAOLA SOMMA AND HUW THOMAS

The purpose of this chapter is to highlight some comparisons based on the case study chapters, and to suggest research agenda.

The composition of ethnic minority populations

The structure of the case studies sought to emphasise three concepts: urban policy frames, power relations within policy processes, and racialisation. The histories of ethnic minority populations, and the racialisation of politics, have played a significant role in shaping the policy frames and also the politics of national and local governments.

Of all the countries featured in this book, only Sweden does not have a colonial past. Racial ideology in the other countries is closely related to colonial subjugation of non-Europeans. Britain stands out as an archetypal coloniser which has integrated values and attitudes of racial superiority into a racial discourse central to all public policy. In France, despite the much-proclaimed 'Republican Model' whereby anyone willing to accept the rules of the polity and adopt the national culture could become a French citizen, race has played a significant role in politics, especially in local politics.

Despite the colonial past of five out of the six countries covered by the case studies, these countries differ with regard to the size and composition of their immigrant population. The share of ethnic minority population in Italy and Portugal is relatively small. For both these countries, it has been emigration, rather than immigration, that was a dominant phenomenon until the 1970s. Britain, on the other hand, has a long history of immigration, partly from European countries - especially Ireland - and partly from Commonwealth countries (see Khakee and Thomas, 1995). So does France, but early immigration to France was mainly from other European countries. The relative share of immigrants, however, is highest in the two north-western European countries: 15.6 per cent of the Dutch population has a non-Dutch background, and 18.6 per cent of Sweden's population is classified as immigrant. The Swedish figures, however, include persons born in Sweden but having at least one parent who is foreign-born. Labour immigration has been pronounced in the case of France, the Netherlands, and Sweden.

In all the six countries, discussions of immigration are framed by the conception of the 'immigrant as problem'. However, there are subtle differences. Black people, in the sense of people of African descent, are targets of racism in all six countries. However, in Britain the Asian population is also a major target; in France the main target is the North African population from the Magreb countries. In the other four countries the ideology and notions of superiority are more diffused and dispersed over various ethnic minorities. In the case of Sweden the term 'blackheads' is derogatorily used, even for southern European minorities. One aspect with regard to the settlement of the immigrant population is its regional concentration. A majority of immigrants live in large urban regions, especially metropolitan areas or conurbations. Moreover there is further concentration at district level: in Britain, France, Italy, and Portugal, in a smaller number of city areas; and in the Netherlands and Sweden, in specific peripheral housing districts. This has resulted in linking spatial location with race (Solomos and Back, 1995). The racist rhetorics often point out that certain city areas are being transformed into 'alien territory', and there is an increasing tendency to link urban policy questions with the law-and-order issue (see, e.g., Solomos, 1993).

Local governance

Local governance in the six European countries represented in our case studies differs so widely that it is difficult to draw conclusions with regard to similar or contrary development tendencies which have relevance for the main subject of this book: namely urban renewal and social exclusion. However, there are certain common trends in the development of local governance which have importance in this context.

With the exception of Great Britain (until the late 1990s), one general trend has been towards increasing decentralisation of state functions, and devolution of power to local government. Specific features of this trend are, for example, the changes in the legislation governing the organisation of municipal governance: the tendency has been to give greater freedom to local governments to choose governance forms; and the replacement of specially 'earmarked' state grants by lump sum grants giving local governments greater freedom to adjust spending to local needs. Moreover, the trend has been towards replacing detailed statutory regulations by framework legislation. However, the state still retains considerable power, both with regard to the level of local governments' taxation, and to the use of public money for different purposes (see, e.g., Goldsmith, 1992). The relevant issue in the case studies has been the impact of this decentralisation on urban policy, especially with regard to the plight of ethnic minorities. The tendency has been an increasing unwillingness to direct programmes for the benefit of minorities, for fear of offending the majority of the electorate.

The second major trend is the changing nature of territorial relationship at the local level. None of the six countries are 'pure' unitary states. During the 1970s and 1980s, regional governance has been introduced over and above the existing structure of local governance. There are several reasons for this development. Progress towards European integration has involved a decline in the role of the national state, while that of local government has been strengthened. The role of urban regions in European economic development has been gradually increasing. Another reason for the promotion of regionalism has been a political desire to create economically viable regions in the face of economic globalisation. A clear tendency in this context has been to increase attempts at inter-municipal co-operation, in order to make effective use of resources, especially physical and social infrastructure. An important feature of this new regionalism has been urban policy directed towards space marketing, with its emphasis on flagship projects. This development has taken place at the cost of urban measures to enhance social integration (see, e.g., Bianchini and Parkinson, 1993).

There is a considerable difference between the six countries with regard to the responsibilities of local government. In Britain, the Netherlands, and Sweden, local government has had a major share in the implementation of the national welfare programme, besides its responsibility for the management of technical amenities and land-use. In

France, Italy, and Portugal, central and local government shares of responsibility with regard to welfare development have been more diffused. In the last few years, local government has had an increasing propensity to take greater responsibility with regard to attempts to stimulate economic development within its territory. Municipalities have been increasingly balancing their traditional responsibility with this new one (see, e.g., Wolman and Goldsmith, 1992). This development has not always been helpful to the less-favoured part of the population. Expenditure on social welfare has been reduced, and specified aid to minority groups has diminished.

A fourth trend has been the changing relationship with the private sector. Urban governments have increasingly delegated the management of public services to the private sector. Contracting out has become increasingly common. The previous boundary between the private and public sectors has been reappraised, and private enterprises have been invited to submit tenders even for activities which belong to the core of the urban public sector (Council of Europe, 1993). The concept of local government as an 'enabling authority', which minimises its own activity and engages private enterprises, has become increasingly accepted by local governments (Lidström, 1996). Some of this 'privatisation' has had a negative impact on the availability of services to the less-favoured parts of the population, mainly because services have become more expensive. The situation for less-favoured minorities has worsened as the fiscal squeeze has led central government to reduce its grants to local government.

Finally, electoral support for local governance has been undergoing significant change. Although the six countries represented in the case studies make use of different electoral systems and have different historical traditions, two aspects seem to be shared by all six countries. Local elections are dominated by the major national parties, which means that their political agenda, to a considerable extent, are dictated by national issues. This has led to increasing party conflict at local level, where previously the tendency had been to get concrete results. The second aspect is that, with the exception of Sweden (where elections to the local council and national parliament take place on the same day), participation in local elections is lower than in national elections. Municipal referenda are increasingly used as a complement to representative democracy, but their importance is quite limited. These two aspects are not always favourable for ethnic minorities. National political conflicts submerge local issues relating to ethnic minorities' specific needs and preferences.

Urban policy frames and social exclusion

One of the major factors affecting urban policy frames has been the emergence of racism as a public issue. The latter has been due to economic and social crises, as a result of the changing composition and size of the immigrant population, and the rise of extreme right-wing political parties. The Swedish and Dutch public discourses on racism in the 1980s and 1990s are an echo of the British debate from the 1960s. In Britain the legislation limiting and controlling immigration has been, in effect, linked in political and public debates to legislation on improving 'race relations' and combating discrimination. We find similar public discourses in France, the Netherlands, and Sweden. The convergence of immigration policy within the European Union, and increasing co-operation and co-ordination, have resulted in vigorous enforcement of immigration control in all the member countries (Mitchell and Russell, 1994). On the other hand, there is a lack of adequate measures to enforce legislation on integration in these countries.

Developments in the labour market have had an impact on urban policy frames. In all six countries, unemployment has increased rapidly among immigrants compared with their native counterparts, reflecting a highly segmented labour market in which ethnic minority workers, apart from occupying low-paid and low-status jobs, are also concentrated in sectors where there is little security of employment. A recent EU study shows that systematic employer-based discrimination in the labour market is prevalent in all member countries. Unemployed and low-status ethnic minority workers have become increasingly dependent on welfare assistance and other kinds of social help. However, cuts in public expenditures not only reduce this assistance but also limit local authorities' scope for initiating programmes which could decrease the negative impact of unemployment.

Urban governments in these six countries often face an important dilemma, namely the extent to which any policy ensuring equal opportunity can be enforced. There are structural constraints. Urban governments are dependent upon electoral support in order to implement such a policy. The situation varies slightly between the six countries, but on the whole, urban governments are not entirely willing to direct programmes which might clearly and specifically benefit the ethnic minority population, for fear of offending the majority of the electorate. Moreover, the poor representation of ethnic minorities in the political process limits their access to policy-making arenas. The growing impact of neoliberal ideology enhance these constraints.

While there is no unequivocal picture with regard to increasing social exclusion in the six countries, the general pattern shows the concentration of ethnic minorities in specific areas, their precarious situation in the labour market, and their increasing reliance on social welfare assistance (though in Britain, where the ethnic minority population is decreasingly a first generation immigrant population, there is evidence of social and spatial mobility among some ethnic groups). At the same time, these countries face social and economic crises, implying that social exclusion does pose a challenge for local governance.

Urban renewal and social exclusion

Urban renewal, with specific reference to social exclusion, involves two different issues. The first is the relationship between urban regeneration and the racialisation of ethnic minority populations. For example, has the presence of immigrants been a major factor in determining renewal policy? Are renewal policies directed exclusively at disadvantaged and racialised ethnic minorities? The case studies do not provide any unequivocal answer to these questions. There is little evidence of even marginal concern for improving the welfare of ethnic minorities in property-led regeneration, as the British experience shows. Where urban renewal has a broad structural emphasis, problems affecting ethnic minorities are taken into consideration, but along with several other concerns.

The second issue is about the actual impact of urban renewal on the exclusion process. Does renewal lead to exclusion, and does it especially affect ethnic minorities? In Britain as well as France, urban renewal has often led to an increase in property prices, forcing ethnic minorities, together with other less-advantaged groups, to move to other housing districts. In Sweden and the Netherlands, ethnic minorities have often been located in high-rise apartment housing in the city suburbs. Most of this was built in the housing programmes of the 1960s and 1970s, and the renewal of these areas has meant further concentration of ethnic minorities in specific housing districts. In Italy and Portugal, ethnic minorities are often concentrated in derelict areas which have been subject to only marginal renewal efforts.

There are obvious divergences in urban renewal policy in western Europe: on the one hand are the more market-oriented urban renewal policies, which only take into consideration property value; and on the other hand 'structural' approaches, which pay attention to social and economic conditions in the area. The latter also involve a more comprehensive approach to remedying the problems of areas subject to renewal. The approach adopted to renewal appears to determine the amount of influence ethnic minorities have over the process. Portugal seems be anomalous in some respects, inasmuch as the situation of ethnic minorities is part of a much wider problem of illegal housing districts, especially in the Lisbon region.

Research agenda

All the case studies contain a specific section discussing an agenda for future research on the question of urban renewal and social exclusion. The proposals put forward in each case study are context-bound, in the sense that specific circumstances and policy assumptions prevailing in each country have determined the issues for further research.

More generally, major research projects in the future should concentrate on three aspects of the dynamics of social exclusion.

Economic exclusion

By this we mean the degree to which economic opportunities are opened up to, or closed off from, ethnic minority residents of areas affected by urban renewal. Urban renewal can involve the rupturing of delicate economic networks which include linkages, informal economic activities, and family or neighbourhood co-operative efforts. These may not have the high profile which often captures the attention of policy makers, but they are, nevertheless, vital to the survival and social integration of ethnic minorities, particularly immigrants (see, e.g., Lemkow, 1994). For example, existing firms may be lost, and be replaced by economic activity for which the local population is not trained or cannot compete effectively. The informal or 'submerged' economy in the area may also be disrupted.

There is considerable case study-based research evidence which demonstrates that simplistic metaphors are misleading ways to describe the economic impact of urban regeneration projects. Phrases such as 'trickle down', which assume some kind of straightforward transmission of economic benefits from property development, or other highly profitable commercial activity, to the economically disadvantaged, have been shown to be inadequate ways of capturing the complexities of the processes actually involved (see, e.g., Docklands Forum/Birbeck College, 1990, on London Docklands). Another set of questions arises when rehousing of residents is considered. For example, when residents of low quality, low cost housing in densely populated inner city areas are moved to newer residences some distance away, as part of an urban renewal programme, does the relocation disrupt the journeys to work of those already in employment? Is the disruption severe? Is the potential disruption even considered by governance agencies promoting the urban renewal project? If so, how? If not, why not?

Exclusion from governance

Urban renewal may affect voting or electoral participation, as well as membership of political parties, popular movements, and immigrants' associations. Some European countries have enacted legislation which allows foreign nationals with permanent residentship to vote in local elections (e.g. Sweden, Belgium). It would be interesting to find out how such legislation may affect the opportunities of minorities to influence decision-making.

As urban renewal proceeds, local people may find it difficult to influence decision-making about areas with which they are familiar: areas they may regard as home; indeed, the complexity of the private-public partnerships which often promote urban renewal schemes, and the variety of different sources of funding for urban renewal projects, may make it difficult for people to identify the origins of key decisions which are affecting their futures.

A burgeoning international research-based literature on urban governance has highlighted:

(i) the increasing variety of configurations of agencies now involved in urban renewal/regeneration activities, many with, at best, tenuous lines of accountability to local residents;

(ii) the complexity of the policy processes involved in many of these new configurations, as, for example, traditional elected governments work alongside single-purpose quasi-public bodies, or voluntary groups.

This dimension gives rise to a number of theoretical issues which have implications for policy. Firstly, how can agencies which are permeable to influence be created? Secondly, are certain policy frames more supportive of such agencies than others? Further, and crucially, does governance which is locally accountable actually influence the kinds of economic benefits which accrue to local populations in general, and to ethnic minorities in particular? Finally, is it possible to speak of a single 'local community' in these areas of ethnic mix, social deprivation and poverty? If not, what are the implications for establishing structures of community governance, and how can community capacities be developed which help all local people have a real say, and stake, in urban renewal projects, and are *inclusive* (promoting solidarity), but also effective?

A lack of social cohesion

Urban renewal often breaks up existing social networks (extended family contacts, clubs, associations, etc.). This is perhaps the most important impact of immigrant residents being forced out of an area following urban renewal. Two common aspects of this are:

(i) the arrival of relatively more affluent, typically white, newcomers in inner city areas which have hitherto been characterised by low property values, poor environmental condition, and a poor, often multi-ethnic, population;

(ii) the transfer (often under less than voluntary conditions), to peripheral housing areas, of poor residents of highly densely-populated, low property-value, inner city residential areas which are the object of urban renewal projects. There is an historical literature on the insensitivity of the governmental processes involved in these movements in the 1960s and 1970s, and the consequent disruption of established communities. There have been case studies of more recent attempts to promote 'bottom-up' urban renewal, with the implication that social integration will be fostered thereby. However, small communities can be extremely conservative; consequently, empowering them within the urban regeneration process may militate against forging new social networks which involve, and encompass, new residents in an area (see, for example, the chapters on the Netherlands, France, and Italy, in this book). Research is needed to investigate whether this possibility can be avoided, and if so, how.

A concentration of population characterised by poverty and limited life-chances often results in high rates of crime, destitution, and early deaths. Such a development leads to communities which are spatially and otherwise outside the mainstream of society.

References

Bianchini, F. and Parkinson, M. (eds), (1993), *Cultural Policy and Urban Regeneration: The Western European Experience*, Manchester University Press, Manchester.

Council of Europe (1993), *The Role of Competitive Tendering in the Efficient Provision of Local Services*, Report No. 49, Council of Local and Regional Authorities of Europe, Strasbourg.

Docklands Forum/Birkbeck College (1990), *Employment in Docklands*, Docklands Forum, London.

Goldsmith, M. (1992), 'The structures of local government' in Mouritzen P.E. (ed.), *Managing Cities in Austerity: Urban Fiscal Stress in Ten Western Countries*, Sage, London.

Khakee, A. and Thomas, H. (1995), 'Ethnic Minorities and the Planning System in Britain and Sweden', *European Planning Studies*, vol. 3, no. 4, pp. 489-510.

Lemkow, L. (1994), 'Workshop report: migrants and the city', in EFILW *European Conference on Migration and the Social Partners*, EFILW, Dublin, pp. 59-60.

Lidström, A. (1996), *Kommunsystem i Europa (Municipal Systems in Europe)*, Publica, Stockholm.

Mitchell, M. and Russell, D. (1994), 'Race, Citizenship and fortress Europe', in Brown, P. and Crompton, R. (eds), *A New Europe? Economic Restructuring and Social Exclusion*, UCL Press, London, pp. 136-156.

Solomos, J. (1993), *Race and Racism in Britain*, 2nd ed., Macmillan, London.

Solomos, J. and Back, L. (1995), *Race, Politics and Social Change*, Routledge, London.

Wolman, H. and Goldsmith, M. (1992), *Urban Politics and Policy*, Blackwell, Oxford.